ALSO FROM
HILL AND WANG

THE CARTOON INTRODUCTION TO STATISTICS

THE CARTOON INTRODUCTION TO STATISTICS

BY GRADY KLEIN AND ALAN DABNEY, Ph.D.

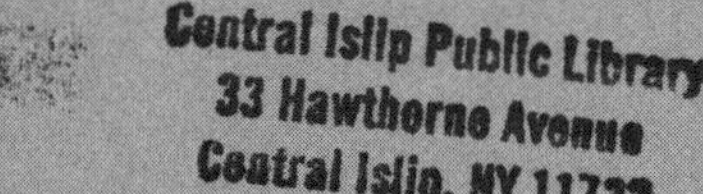

A NOVEL GRAPHIC FROM HILL AND WANG
A DIVISION OF FARRAR, STRAUS AND GIROUX
NEW YORK

HILL AND WANG
A DIVISION OF FARRAR, STRAUS AND GIROUX
18 WEST 18TH STREET, NEW YORK 10011

PRINTED IN THE UNITED STATES OF AMERICA
PUBLISHED SIMULTANEOUSLY IN HARDCOVER AND PAPERBACK
FIRST EDITION, 2013

LIBRARY OF CONGRESS CATALOGING-IN-PUBLICATION DATA
KLEIN, GRADY.
THE CARTOON INTRODUCTION TO STATISTICS / BY GRADY KLEIN AND ALAN DABNEY, PH.D. — FIRST EDITION.
P. CM.
ISBN 978-0-8090-3366-9 (HARDCOVER) — ISBN 978-0-8090-3359-1 (TRADE PBK.: ALK. PAPER)
1. MATHEMATICAL STATISTICS — COMIC BOOKS, STRIPS, ETC. 2. GRAPHIC NOVELS.
I. DABNEY, ALAN, 1976– II. TITLE.

QA276 .K544 2013
519.5 — DC23

2012030027

WWW.FSGBOOKS.COM
WWW.TWITTER.COM/FSGBOOKS • WWW.FACEBOOK.COM/FSGBOOKS

1 3 5 7 9 10 8 6 4 2

FOR ANNE AND LIAM AND BENJAMIN
—GK

FOR ELLIOTT AND LOUISE AND NICK
—AD

CONTENTS

INTRODUCTION: THEY'RE EVERYWHERE ... 1

PART ONE: **GATHERING STATISTICS** ... 15

1. NUMBERS ... 17
2. RANDOM RAW DATA ... 25
3. SORTING ... 39
4. DETECTIVE WORK ... 51
5. MONSTER MISTAKES ... 67
6. FROM SAMPLES TO POPULATIONS ... 81

PART TWO: **HUNTING PARAMETERS** ... 89

7. THE CENTRAL LIMIT THEOREM ... 91
8. PROBABILITIES ... 105
9. INFERENCE ... 121
10. CONFIDENCE ... 131
11. THEY HATE US ... 143
12. HYPOTHESIS TESTING ... 161
13. SMACKDOWN ... 175
14. FLYING PIGS, DROOLING ALIENS, AND FIRECRACKERS ... 191

CONCLUSION: THINKING LIKE A STATISTICIAN ... 205

APPENDIX: THE MATH CAVE ... 213

INTRODUCTION
THEY'RE EVERYWHERE

MOST OF US ENCOUNTER STATISTICS **EVERY DAY**...
AWESOME!
ONE BOWL OF CHOCOLATY FROSTO BOMBS HAS **1,200%** OF MY DAILY RECOMMENDED SUGAR.
...EVEN IF WE DON'T **CRUNCH NUMBERS** FOR A LIVING.

STATISTICS RADIATE FROM OUR **TELEVISIONS**...
THIS SHOW HAS AN **ESTIMATED AUDIENCE OF 4.8 MILLION!**
IT **MUST** BE GOOD.

...EMANATE FROM OUR **PHONES**...
YOU SENT MORE **TEXT MESSAGES** THIS MONTH THAN THE **ENTIRE NATION OF CHAD.**

...SEEP FROM OUR **RADIOS**...
POLLS SHOW SENATOR NEERDORPH WITH A **40-POINT** LEAD!
...AND LITTER OUR **ROADS.**

KILLED OR INJURED?
Average Award $150 million, and Beyond!
Eaton Cake, Attorney at Law
800 WIN CAKE

78%
of Dentists Recommend It!
SCOURADE!
IT'S **IMPOSSIBLE TO ESCAPE THEM.**

THEY'RE **EVERYWHERE!**

...AND LIKE IT OR NOT, WE WILL **BECOME STATISTICS** WHEN WE DIE.

FORTUNATELY, THERE'S A **GOOD REASON** FOR ALL THIS.

STATISTICS ARE EVERYWHERE
BECAUSE THEY'RE SO USEFUL.

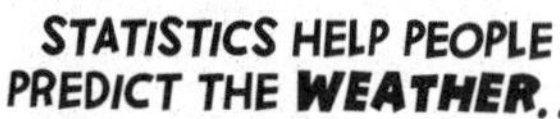
STATISTICS HELP PEOPLE PREDICT THE WEATHER...

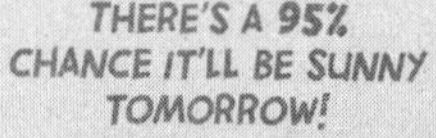
THERE'S A 95% CHANCE IT'LL BE SUNNY TOMORROW!

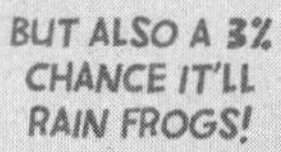
BUT ALSO A 3% CHANCE IT'LL RAIN FROGS!

...AND ORGANIZE THE INTERNET...

BASED ON YOUR BUYING HISTORY, I'VE GOT RECOMMENDATIONS FOR YOU.

HOW DID IT KNOW I WANTED A WILLIAM SHATNER DOLL?

...AND DEVELOP MEDICINES...
OUR STUDIES SHOW THAT THIS PILL IS ONLY 2.5% MORE LIKELY THAN A PLACEBO TO PREVENT CANCER, WITH A MARGIN OF ERROR OF 12%...

...BUT IT WORKS GREAT AS A LAXATIVE!

GREAT, WHAT SHOULD WE CALL IT?

...AND INFLUENCE FASHION.
I USED STATISTICS TO DETERMINE THAT DENIM JACKETS ARE LIKELY TO COME BACK IN STYLE THIS YEAR...

LIKE, WOW YOU ARE SO 1987, I LOVE IT!
BUT LOSE THE BELL-BOTTOMS.

AND THAT'S NOT ALL.

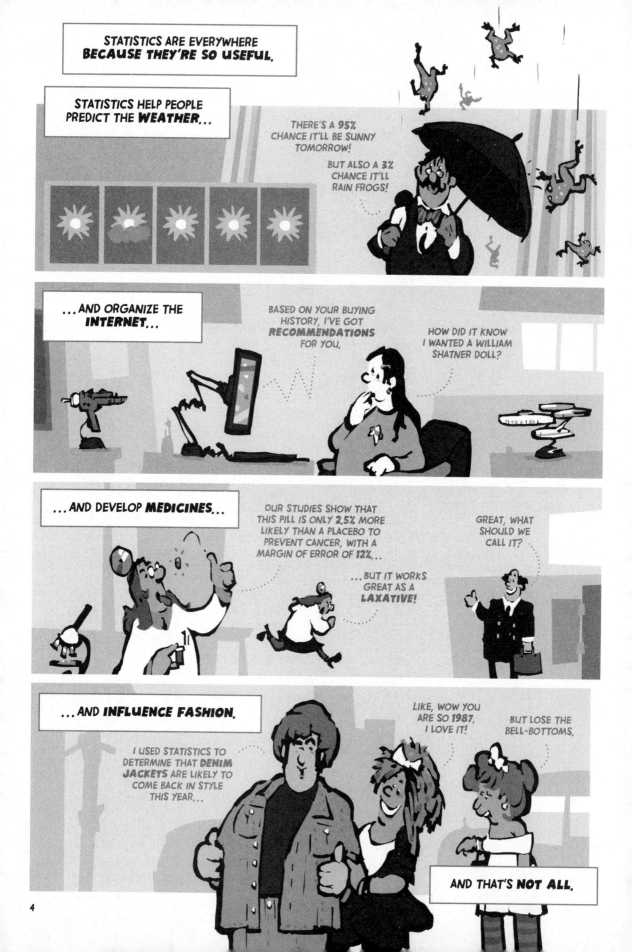

...AND **MAKE MONEY**...

IF CURRENT MARKET TRENDS HOLD...

...I'LL BE BETWEEN 12% AND 15% MORE FILTHY RICH TOMORROW!

...AND **SHOW THEIR SUPERIORITY**.

HA, I HIT MORE HOME RUNS THAN YOU DID.

HA, YOU USED STEROIDS, AND HERE ARE THE STATISTICS TO BACK ME UP!

SO **WHAT MAKES** STATISTICS **SO INCREDIBLY USEFUL?**

THE **SIMPLE ANSWER** IS THAT STATISTICS HELP US COME TO GRIPS WITH **LARGE NUMBERS OF IMPORTANT THINGS...**

...WHICH IN TURN CAN HELP US **BETTER UNDERSTAND OUR COMPLEX WORLD...**

...AND **MANIPULATE IT.**

BUT THE **REAL POWER** OF STATISTICS IS **MORE SPECIFIC.**

HERE'S THE **REAL REASON** EVERYONE USES STATISTICS:

...OF **ALL THE FISH IN A LAKE.**

IF WE FIND OUT HOW MUCH EACH FISH WEIGHS ON AVERAGE...

...WE CAN FIGURE OUT ABOUT HOW MANY FISH WE NEED TO CATCH EACH DAY TO **KEEP OUR CATS FROM STARVING!**

IF WE **DRAINED THE LAKE** AND **WEIGHED EVERY SINGLE FISH**...

...WE'D HAVE **ALL THE INFORMATION** WE NEED TO CALCULATE THE ANSWER.

BUT FOR OBVIOUS REASONS, WE **CAN'T DO THAT.**

ON THE OTHER HAND, IF WE CATCH A **SAMPLE** OF **100 FISH** AND WEIGH THEM...

THESE **100** FISH WEIGH **247** POUNDS.

SO THE **AVERAGE FISH IN THIS SAMPLE** WEIGHS **2.47** POUNDS!

...WE'LL HAVE ONLY **LIMITED INFORMATION** ABOUT ALL THE FISH.

SO NOW WE KNOW THE AVERAGE WEIGHT **IN THIS SAMPLE PILE**...

...BUT WE STILL DON'T KNOW THE AVERAGE WEIGHT OF THE **REST OF THE FISH IN THE LAKE**.

BUT HERE'S THE **COOL THING:**

WITH THE TOOLS OF STATISTICS WE CAN **USE THIS LIMITED INFORMATION**...

...TO MAKE **CONFIDENT STATEMENTS** ABOUT **ALL** THE FISH IN THE LAKE.

STATISTICS IS ABOUT **USING THE FISH WE DID CATCH**...

...TO **SAY THINGS ABOUT THE FISH WE DIDN'T**.

THIS BOOK IS ABOUT THE **FUNDAMENTAL QUESTION** OF STATISTICS:
HOW DO WE USE **SAMPLES**...

...TO MAKE **CONFIDENT STATEMENTS** ABOUT **ENTIRE POPULATIONS?**
ALL STATISTICS PROBLEMS LOOK LIKE THIS!

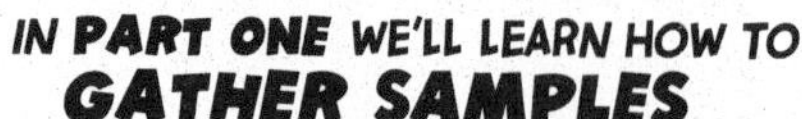

THEN, IN **PART TWO,** WE'LL LEARN HOW TO USE SAMPLES TO
HUNT FOR QUALITIES IN POPULATIONS...

...USING A PROCESS CALLED **STATISTICAL INFERENCE.**

WHAT CAN
THESE FISH...

...TELL US ABOUT
THOSE FISH?

ALONG THE WAY WE'LL LEARN TO SIFT THROUGH **BIG PILES OF DATA**...

...CALCULATE **CONFIDENCE INTERVALS**...

I'M 95% CONFIDENT THAT WE HATE YOU ABOUT THIS MUCH.

...AND **TEST HYPOTHESES.**

I'M 3% CONFIDENT THAT MY EVIL MACHINE **ISN'T BUSTED!**

AND MORE GENERALLY, WE'LL GET A SENSE OF THE KIND OF THINGS YOU **CAN**...

...AND **CAN'T**...

...DO WITH STATISTICS.

WE CAN USE STATISTICS TO **MAKE CONFIDENT GUESSES**...

...BUT YOU CAN **NEVER** USE THEM TO **ACHIEVE CERTAINTY.**

IF WE DON'T CATCH **ALL** THE FISH...

...WE'LL NEVER KNOW **FOR CERTAIN** WHAT'S DOWN THERE.

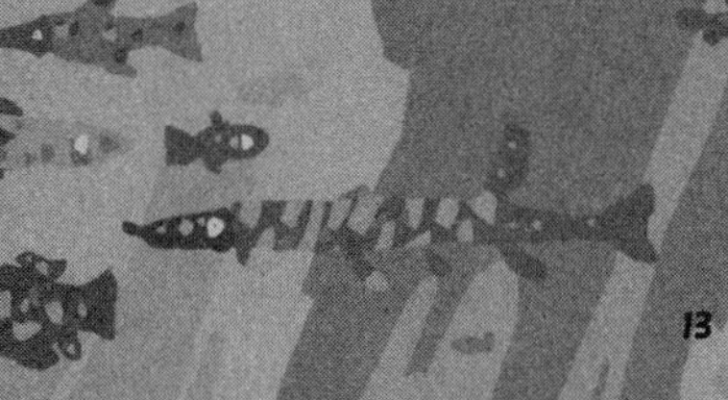

IN THIS BOOK WE'RE GOING TO **FOCUS ON THE BASIC CONCEPTS.**

LIKE **STANDARD DEVIATIONS...**

...AND **SAMPLING DISTRIBUTIONS...**

...AND **PROBABILITIES...**

...AND **CONFIDENCE!**

BUT IF YOU'RE ALSO CURIOUS ABOUT THE **TECHNICAL DETAILS...**

LIKE WHAT THE HECK DO THESE **FORMULAS** AND **SYMBOLS** MEAN?

...YOU CAN FIND THOSE IN A SECTION AT THE END CALLED **THE MATH CAVE.**

PART ONE
GATHERING STATISTICS

CHAPTER 1
NUMBERS

IN THIS CORNER, WEIGHING IN AT
50.8 TRILLION NANOGRAMS...
...THE DWARF!
AND IN THIS CORNER, WEIGHING IN AT
0.193 TONS...
...THE GIANT!

NEVERTHELESS, STATISTICS **DOES** INVOLVE **WRESTLING WITH NUMBERS**...

SOME NUMBERS ARE **BIG**...

THERE ARE 10^{14} POSSIBLE NERVE CONNECTIONS IN YOUR BRAIN!

THAT'S 100,000,000,000,000!

...AND SOME ARE **SMALL**.

EACH ATOM IS 10^{-17} AS LARGE AS YOUR EYEBALL.

THAT'S 100,000,000,000,000,000 TIMES SMALLER!

BUT SOME **BIG** NUMBERS DESCRIBE **SMALL** THINGS...

THERE ARE MORE THAN **500 BILLION INSECTS** LIVING IN TEXAS.

...AND VICE VERSA.

THERE IS EXACTLY **ONE STAR IN THE SOLAR SYSTEM**.

SOME NUMBERS ARE **POSITIVE**...

WE WON BY **TWO GOALS!**

...AND SOME ARE **NEGATIVE**.

THE DOW JONES IS **DOWN 423 POINTS TODAY**. URGH!

BUT SOME **POSITIVE** NUMBERS DESCRIBE **NEGATIVE** THINGS...

UNEMPLOYMENT **INCREASED 4%** LAST MONTH.

...AND VICE VERSA.

THE NUMBER OF MURDERS IN THE CITY HAS **SHOT DOWN**.

AND **THAT'S NOT ALL**...

SOME NUMBERS **SEEM SCARY**...

YOU HAVE MORE THAN **TWO POUNDS** OF BACTERIA LIVING INSIDE YOU!

...OTHERS **SEEM REASSURING.**

IF THEY'VE SOLD MORE THAN A **BILLION** HAMBURGERS...

...THEY **MUST BE GOOD!**

SOME **REVEAL A GREAT DEAL...**

WE'VE ERADICATED **99.99%** OF THE **SMALLPOX VIRUS** WORLDWIDE.

ALL THESE **FACTS ABOUT NUMBERS** MAKE IT **EASY TO USE THEM**...

...TO **LIE.**

SADLY, THIS CAN MAKE PEOPLE **OVERLY SUSPICIOUS ABOUT NUMBERS GENERALLY**...

...AND **IGNORANT** OF THEIR **VERY REAL POWER.**

THE SOLUTION TO THIS PROBLEM...
SOME NUMBERS ARE TRUE...
...BUT YOU MUST RESIST THE TEMPTATION TO BELIEVE THE FALSE ONES!
HOW DO I KNOW WHICH IS WHICH?
...IS TO TREAT ALL NUMBERS...

...NO MATTER HOW LARGE...
...OR SMALL...
...OR SLEEP INDUCING...
ZZZZZ
...WITH A HEALTHY DOSE OF SKEPTICISM.

THESE COOKIES ARE 100% ORGANIC, AND 98.3% VEGAN...
...BUT THEIR TRANS-ISOMER LACTO-OVO QUOTIENT IS WELL WITHIN THE RECOMMENDED HOMOCYSTEINE RATIO.
I'LL TAKE ONE WITH A GRAIN OF SALT, PLEASE.
THAT'S THE FIRST LESSON OF THIS BOOK.

HELP!
I'M UNCERTAIN ABOUT THESE NUMBERS.
GOOD!
LEARN TO ENJOY IT!
IN STATISTICS, YOU'RE SUPPOSED TO FEEL THAT WAY.

AS WE'LL LEARN LATER, THE **TOOLS OF STATISTICS** CAN HELP US **PUT NUMBERS IN PERSPECTIVE.**

THERE ARE AT LEAST **100 BILLION** BLACK HOLES IN THE UNIVERSE!

I KNOW YOU DIDN'T **LOOK AT THEM ALL...**

...SO HOW **CONFIDENT** ARE YOU ABOUT THAT?

BUT FOR NOW, JUST REMEMBER THAT **PEOPLE ALWAYS HAVE REASONS** FOR CRUNCHING NUMBERS...

THESE NUMBERS **PROVE THAT I'M RIGHT!**

THESE OTHER NUMBERS **PROVE THAT YOU'RE WRONG!**

...AND IT NEVER HURTS TO WONDER **WHAT THOSE REASONS ARE.**

...YOU SHOULD **ALWAYS ASK THESE QUESTIONS** OF **ANY NUMBER YOU ENCOUNTER:**

CHAPTER 2
RANDOM RAW DATA

SINCE THE **DAWN OF CIVILIZATION...**

... PEOPLE HAVE HAD THE **URGE TO COUNT THINGS.**

IN FACT, THE EARLIEST **FORMS OF WRITING** WERE **INVENTED FOR COUNTING.**

AS CIVILIZATIONS **GREW...**

...SO DID THE **NUMBER OF THINGS** NEEDING TO BE COUNTED.

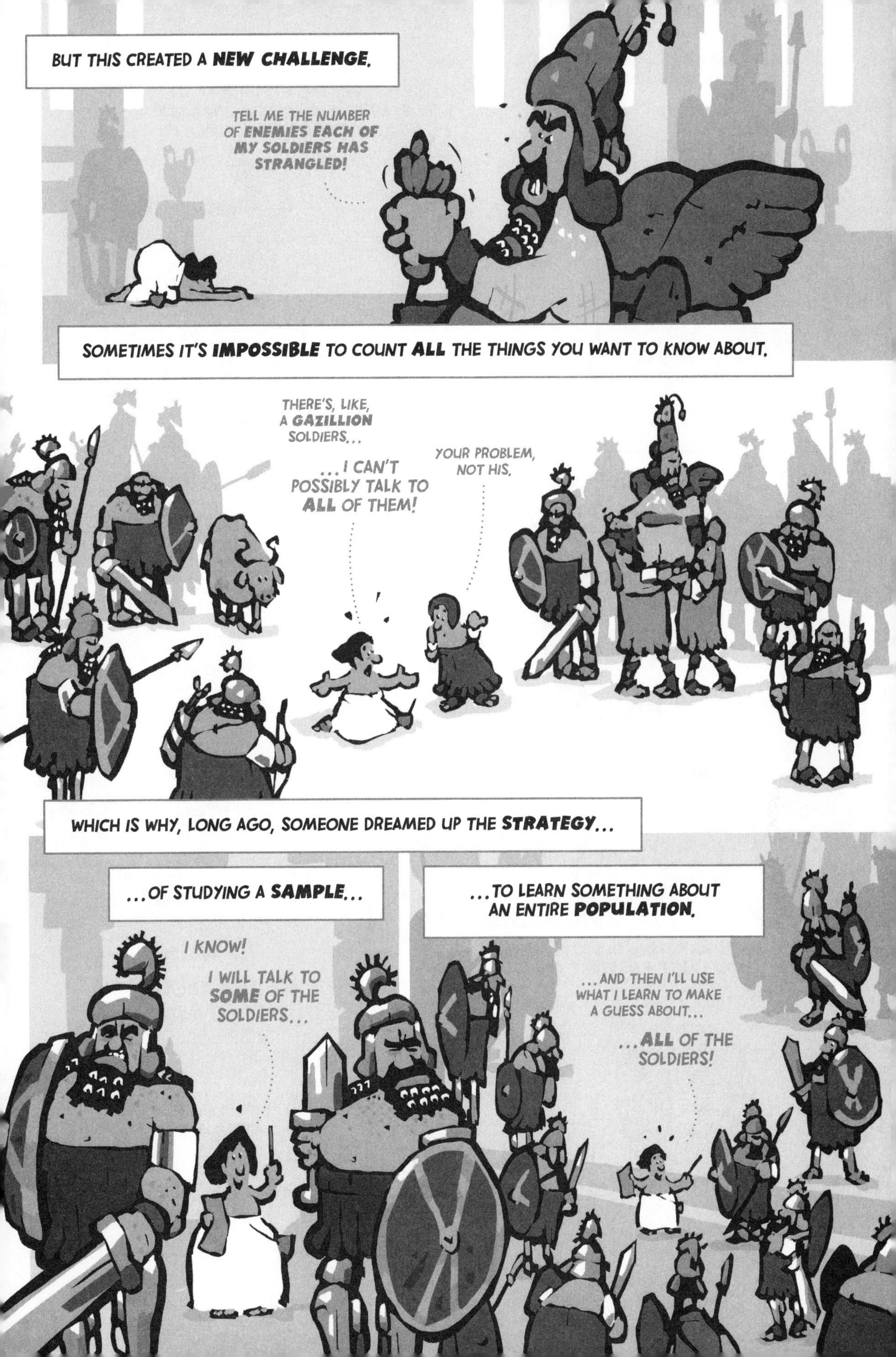

BUT THIS CREATED A **NEW CHALLENGE.**

TELL ME THE NUMBER OF **ENEMIES EACH OF MY SOLDIERS HAS STRANGLED!**

SOMETIMES IT'S **IMPOSSIBLE** TO COUNT **ALL** THE THINGS YOU WANT TO KNOW ABOUT.

THERE'S, LIKE, A **GAZILLION** SOLDIERS...

...I CAN'T POSSIBLY TALK TO **ALL** OF THEM!

YOUR PROBLEM, NOT HIS.

WHICH IS WHY, LONG AGO, SOMEONE DREAMED UP THE **STRATEGY**...

...OF STUDYING A **SAMPLE**...

...TO LEARN SOMETHING ABOUT AN ENTIRE **POPULATION.**

I KNOW!

I WILL TALK TO **SOME** OF THE SOLDIERS...

...AND THEN I'LL USE WHAT I LEARN TO MAKE A GUESS ABOUT...

...**ALL** OF THE SOLDIERS!

USING SAMPLES TO DESCRIBE POPULATIONS IS **CLEVER...**

...BUT THERE ARE A FEW FACTS TO KEEP IN MIND BEFORE WE **ACTUALLY TRY TO START DOING IT.**

FIRST, IT'S **IMPOSSIBLE** TO USE A SAMPLE TO ACHIEVE **ABSOLUTE CERTAINTY** ABOUT A POPULATION.

THAT'S WHY STATISTICS IS ABOUT MAKING OUR **BEST POSSIBLE GUESS...**

...AND **NEVER** ABOUT **BEING CERTAIN.**

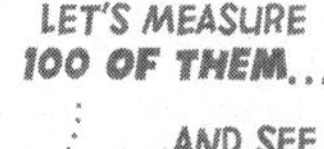

SECOND, IF WE'RE **STUCK WITH A SINGLE SAMPLE**...

I CAN LEARN TO SAY SOMETHING ABOUT **ALL THE SQUID IN THE OCEAN**...

...BY STUDYING **ONLY THESE 35!**

...WE'D BETTER BE SURE WE **COLLECTED IT CAREFULLY!**

UM, DID YOU **WASH YOUR HANDS** BEFORE YOU PICKED THOSE UP?

BECAUSE **ANY MISTAKES** WE MAKE WHEN WE COLLECT OUR SAMPLE...

...CAN **TOTALLY SCREW UP** WHAT WE CONCLUDE ABOUT THE LARGER POPULATION.

NOWADAYS, PEOPLE **COLLECT SAMPLE OBSERVATIONS** IN ***MANY DIFFERENT WAYS...***

ASKING

COUNTING

WEIGHING

MEASURING

BITING

...AND IT'S **NOT EASY WORK.**

IT HELPS TO BE **DETAIL ORIENTED.**

TAKING GOOD MEASUREMENTS CAN BE PARTICULARLY DIFFICULT AT A **LARGE**...

USE **THIS RULER**...

...TO MEASURE THOSE **DRAGONS**...

...OR A **SMALL SCALE**...

...AND THESE **DRAGONFLIES.**

...OR WHEN **MULTIPLE OBSERVERS** ARE INVOLVED.

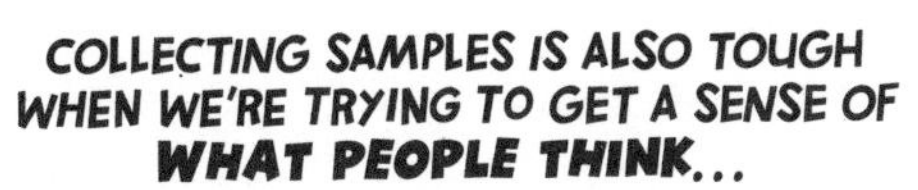
COLLECTING SAMPLES IS ALSO TOUGH WHEN WE'RE TRYING TO GET A SENSE OF **WHAT PEOPLE THINK...**
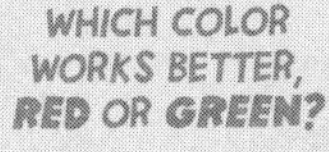
WHICH COLOR WORKS BETTER, **RED** OR **GREEN?**

I'M **COLOR-BLIND.**
...OR **HOW THEY FEEL...**

HOW MUCH DO YOU **LOVE YOUR NEIGHBORS?**

I **HATE 'EM!**
...OR ANYTHING ELSE THEY **MIGHT NOT WANT TO TALK ABOUT...**

HOW MANY **BANKS HAVE YOU ROBBED?**

DEPENDS **WHO'S ASKING.**
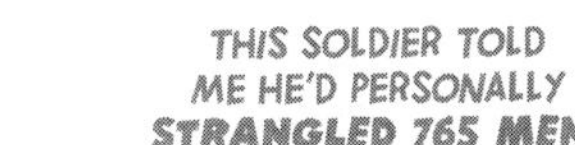
...OR **MIGHT WANT TO EXAGGERATE.**

THIS SOLDIER TOLD ME HE'D PERSONALLY **STRANGLED 765 MEN!**

AND YOU **BELIEVED** HIM?

...IS FIGURING OUT EXACTLY **WHAT TO INCLUDE IN IT.**

THE GOAL IS TO AVOID ANY **BIAS IN OUR SAMPLE...**

...THAT MIGHT LEAD US TO **MISCHARACTERIZE THE POPULATION.**

IF YOU INTERVIEW TOO MANY **POLITE** SOLDIERS...

...YOU'RE GOING TO THINK THE ARMY IS **NICER THAN IT REALLY IS!**

IF YOU INTERVIEW TOO MANY **TERRIFYING** SOLDIERS...

...YOU'RE GOING TO THINK THE ARMY IS **SCARIER THAN IT REALLY IS!**

IDEALLY, WE'D LIKE TO GATHER A SAMPLE THAT **ACCURATELY MIRRORS THE POPULATION.**

THIS CAN SEEM LIKE AN **IMPOSSIBLE TASK...**

...BUT STATISTICIANS HAVE CONCOCTED **A RELIABLE WAY AROUND IT.**

TO AVOID BIAS, WE **ALWAYS** COLLECT SAMPLES **RANDOMLY.**

COLLECTING RANDOM SAMPLES IS A SIMPLE IDEA...

LET'S PUT THE ENTIRE ARMY IN THIS HELMET...
...AND RANDOMLY PULL OUT ONE SOLDIER AT A TIME.

IT'S LIKE BINGO!

...THAT CAN BE HARD TO PULL OFF.
CLIMB IN, BOYS!

IN PRACTICE, WE OFTEN HAVE TO IMAGINE ALL THE THINGS THAT MIGHT BIAS OUR SAMPLE...

I DON'T WANT TO INTERVIEW TOO MANY SOLDIERS WHO ARE ASLEEP IN THE GUTTER...
...OR WHO STINK...
...OR WHO LIVE IN THIS CITY...

...AND MAKE SURE THEY DON'T.

SO I'LL CLOSE MY EYES...
...AND HOLD MY NOSE...
...AND WANDER AROUND THE ENTIRE EMPIRE...

...AND INTERVIEW THE SOLDIERS I RANDOMLY BUMP INTO.

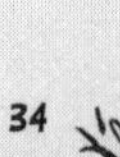

OF COURSE, EVEN WHEN WE COLLECT OUR SAMPLE **RANDOMLY**...

...WE STILL CAN'T **GUARANTEE** THAT IT WILL **ACCURATELY MIRROR** THE POPULATION IT COMES FROM.

IN FACT, ANY ONE **RANDOM SAMPLE**...

...WILL **PROBABLY LOOK DIFFERENT** FROM **THE OVERALL POPULATION**...

...AND FROM **ANY OTHER** RANDOM SAMPLE WE MIGHT RANDOMLY COLLECT.

THE REASON RANDOM SAMPLING WORKS SO WELL IS THAT IT MEANS WE'RE **JUST AS LIKELY** TO GRAB ANY ONE SAMPLE...

...AS ANY OTHER...

...AND **IF** THEY'RE DIFFERENT...

...IT'S **ONLY** BECAUSE OF **CHANCE.**

...BUT IT'S **INCREDIBLY IMPORTANT** TO GET IT RIGHT...

...BECAUSE **RANDOM SAMPLING*** IS THE **KEY** TO ALL STATISTICAL INQUIRY.

* SEE **PAGE 214** FOR SOME TECHNICAL DETAILS.

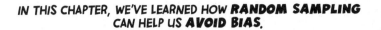

IN THIS CHAPTER, WE'VE LEARNED HOW **RANDOM SAMPLING** CAN HELP US **AVOID BIAS.**

I COULD BE **SCREWING UP MY IMAGE OF THE ARMY...**
...BY INTERVIEWING ONLY **SOLDIERS WHO WON'T KILL ME.**

BUT RANDOM SAMPLES ARE **ALSO** A VITAL PART OF THE **STATISTICAL MACHINERY** WE'LL BE LEARNING ABOUT LATER.

ALL THE TOOLS WE'LL LEARN ABOUT IN PART TWO **REQUIRE RANDOM SAMPLES.**

...ADJUST THIS KNOB...

IF YOUR SAMPLE **ISN'T RANDOM...**
...THE ONLY THING THAT POPS OUT IS **GOBBLEDYGOOK!**

A PILE OF COLLECTED OBSERVATIONS IS CALLED **RAW DATA.**
NOW ALL WE HAVE TO DO IS **COOK IT!**
SINCE THE EARLY DAYS OF CIVILIZATION, THE AMOUNT OF RAW DATA IN THE WORLD KEEPS **INCREASING**...
DO YOU THINK THERE'S **SOMETHING ELSE** WE COULD USE TO WRITE ON?
...AND **INCREASING**...
SORRY, THE LIBRARY OF ALEXANDRIA IS **FULL**...
...LET'S BURN IT DOWN AND START OVER!
...AND **INCREASING**...
WE'RE **GOOGLE!**
...AND **INCREASING**...
WE ARE NO LONGER MERELY **GOOGLE**...
...WE ARE NOW **GOOGLE SQUARED!**
...BUT THE GOAL OF STATISTICS **REMAINS THE SAME.**
WE LOOK AT **RANDOM SAMPLES**...
...AND USE THEM TO MAKE **GUESSES ABOUT THE POPULATIONS THEY COME FROM.**

CHAPTER 3
SORTING

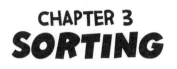

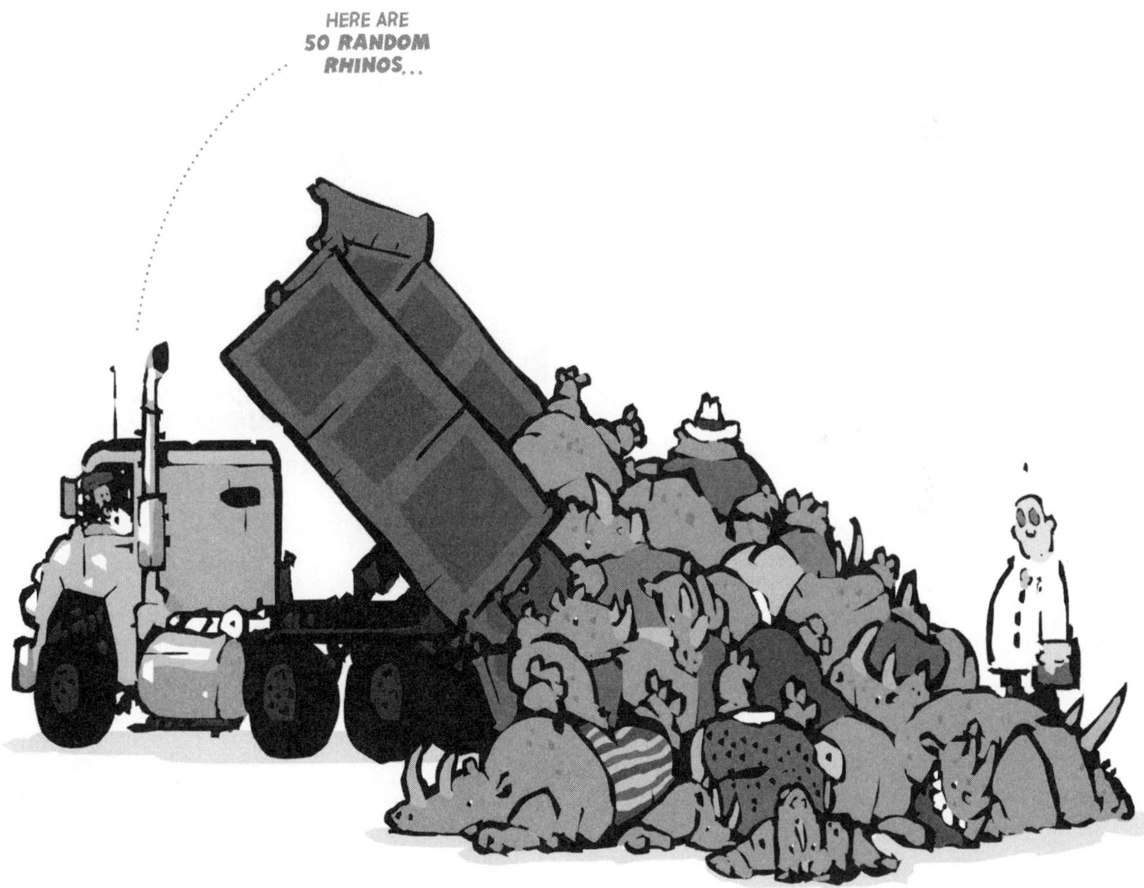
HERE ARE
50 RANDOM
RHINOS...

WE ONLY BOTHER TO COLLECT A RANDOM **SAMPLE**...
...WHEN WE'RE **CURIOUS TO KNOW SOMETHING** ABOUT THE LARGER POPULATION IT COMES FROM.

WELL... **WHAT DO YOU WANT TO KNOW** ABOUT US?

SOMETIMES WE'RE CURIOUS ABOUT **QUALITIES** OR **CATEGORIES**...
...AND SOMETIMES WE'RE CURIOUS ABOUT QUESTIONS WE CAN ANSWER WITH **NUMBERS**.

DO YOU **USE DEODORANT?**
WHERE WERE YOU **BORN?**
WHAT'S YOUR **FAVORITE FOOD?**
HOW MUCH DO YOU **SLEEP?**
HOW THICK IS **YOUR SKIN?**
HOW OFTEN DO YOU **BATHE?**

...AND THE TWO KINDS OF DATA **DON'T MIX**.

WE GATHER **CATEGORICAL DATA...**

...WHEN WE'RE STUDYING FEATURES THAT WE CAN **DESCRIBE ONLY WITH WORDS...**

...OR **YES/NO ANSWERS.**

AFTER WE GATHER CATEGORICAL DATA WE CAN EASILY **PILE IT...**

...OR **SLICE IT...**

...TO GIVE US A SENSE OF THE **PROPORTIONS** IN OUR SAMPLE.

WE GATHER **NUMERICAL DATA...**

...WHEN WE'RE STUDYING FEATURES THAT WE CAN **COMPARE USING NUMBERS.**

AS WE'LL SEE IN PART TWO, ALL THESE NUMBERS MAKE NUMERICAL DATA **MUCH MORE USEFUL** OVERALL.

THE **CRUCIAL DIFFERENCE** BETWEEN THE TWO TYPES OF DATA...

...IS THAT **WE CAN'T DO MATH ON CATEGORICAL DATA...**

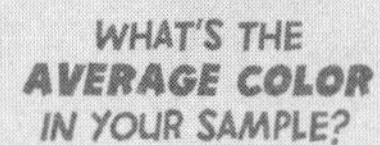

...BUT WE **CAN DO MATH ON NUMERICAL DATA!**

WHAT'S THE **AVERAGE LENGTH** IN YOUR SAMPLE?

0.004 METERS ON THE NOSE!

...BUT **INTIMIDATING TO NORMAL PEOPLE.**

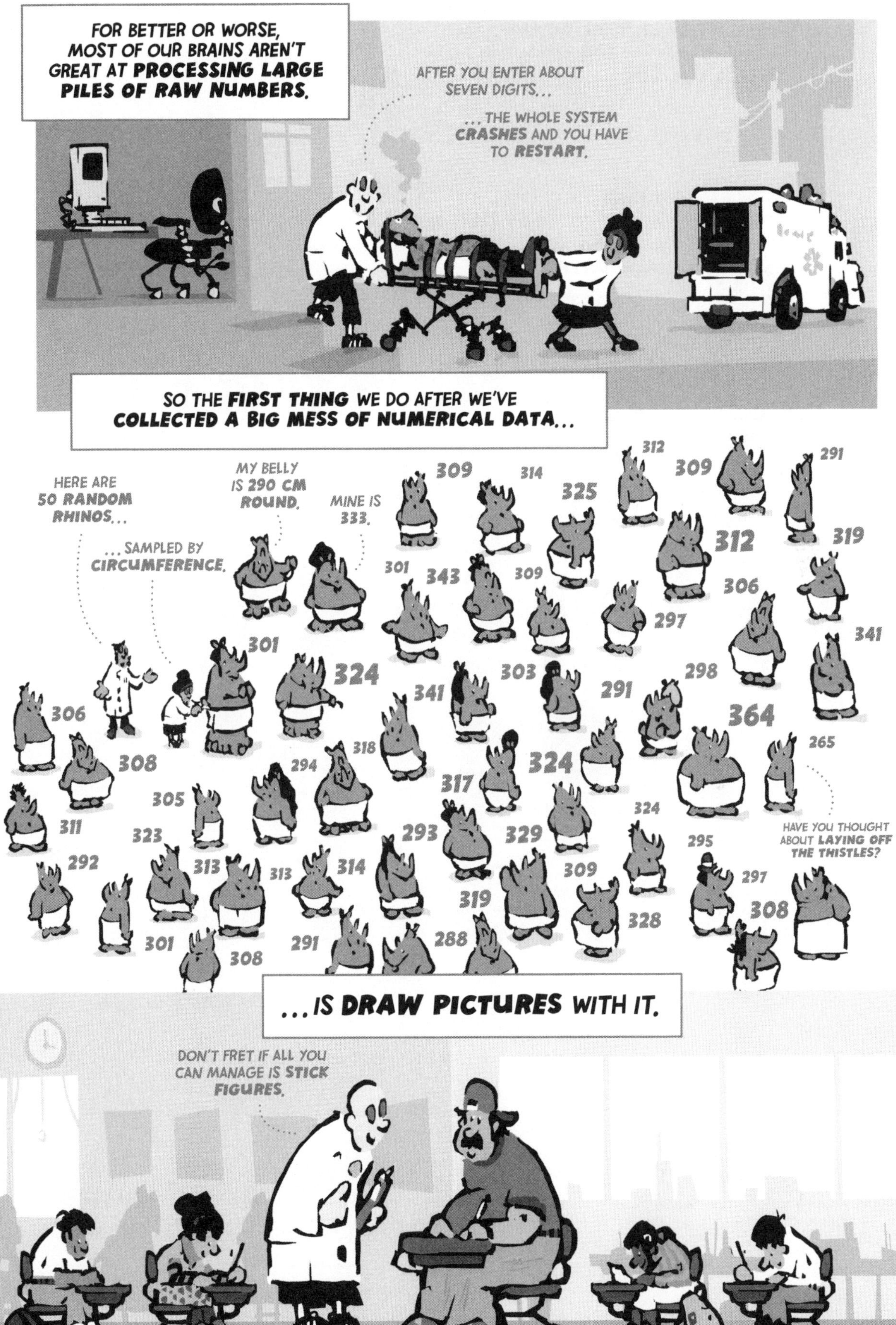

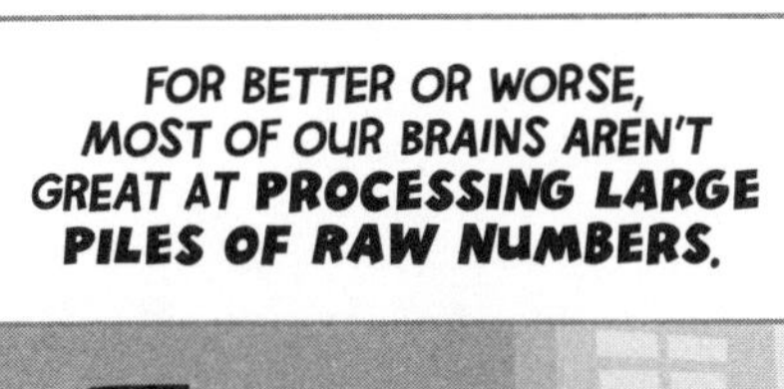

FOR BETTER OR WORSE, MOST OF OUR BRAINS AREN'T GREAT AT **PROCESSING LARGE PILES OF RAW NUMBERS.**

AFTER YOU ENTER ABOUT SEVEN DIGITS...
...THE WHOLE SYSTEM **CRASHES** AND YOU HAVE TO **RESTART.**

SO THE **FIRST THING** WE DO AFTER WE'VE **COLLECTED A BIG MESS OF NUMERICAL DATA...**

HERE ARE **50 RANDOM RHINOS...**

...SAMPLED BY **CIRCUMFERENCE.**

MY BELLY IS **290 CM ROUND.**

MINE IS 333.

309 314 312 309 291

325 312 319

301 343 309 306

297 341

301 324 303 291 298

306 341 364

308 318 294 324 265

305 317

311 323 293 329 324

292 313 313 314 309 295 297

319 328 308

301 291 288 308

HAVE YOU THOUGHT ABOUT **LAYING OFF THE THISTLES?**

...IS **DRAW PICTURES** WITH IT.

DON'T FRET IF ALL YOU CAN MANAGE IS **STICK FIGURES.**

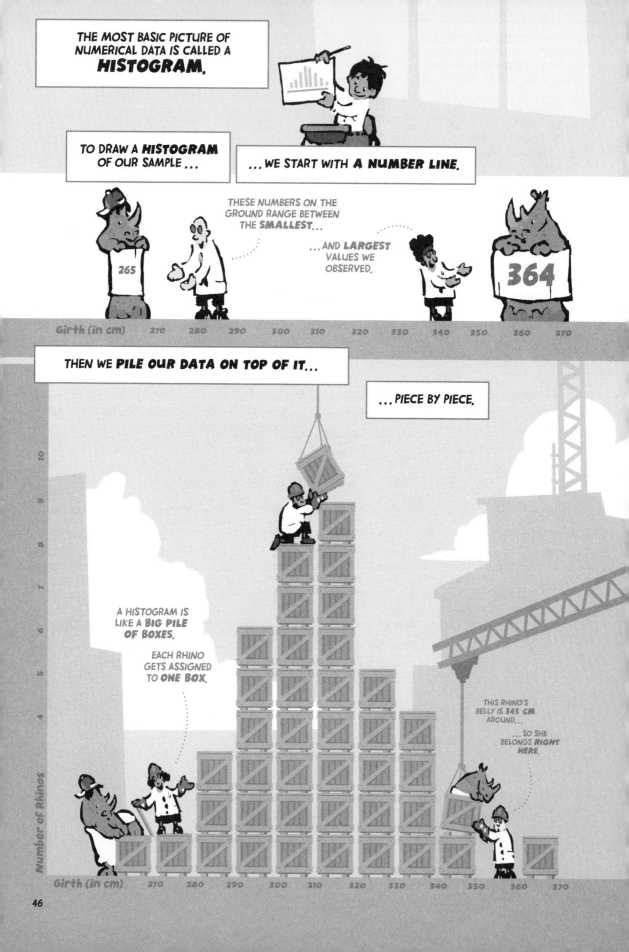

THE MOST BASIC PICTURE OF NUMERICAL DATA IS CALLED A **HISTOGRAM.**

TO DRAW A **HISTOGRAM** OF OUR SAMPLE . . .

. . . WE START WITH **A NUMBER LINE.**

265
THESE NUMBERS ON THE GROUND RANGE BETWEEN THE **SMALLEST** . . .

. . . AND **LARGEST** VALUES WE OBSERVED.

364

Girth (in cm) 270 280 290 300 310 320 330 340 350 360 370

THEN WE **PILE OUR DATA ON TOP OF IT . . .**

. . . PIECE BY PIECE.

A HISTOGRAM IS LIKE A **BIG PILE OF BOXES.**

EACH RHINO GETS ASSIGNED TO **ONE BOX.**

THIS RHINO'S BELLY IS **343 CM** AROUND . . .

. . . SO SHE BELONGS **RIGHT HERE.**

Number of Rhinos 4 5 6 7 8 9 10

Girth (in cm) 270 280 290 300 310 320 330 340 350 360 370

ANOTHER USEFUL WAY TO VISUALIZE NUMERICAL DATA IS WITH A **BOXPLOT.**

TO DRAW A **BOXPLOT** OF OUR SAMPLE...

...WE START WITH **THE SAME NUMBER LINE...**

I'M THE **SMALLEST VALUE** IN THE ENTIRE SAMPLE.
265

I'M THE **LARGEST VALUE** IN THE ENTIRE SAMPLE.
364

Girth (in cm) 270 280 290 300 310 320 330 340 350 360 370

...BUT IN THIS CASE WE **CRAM THE MIDDLE 50%** OF OUR SAMPLE VALUES INTO **ONE BIG BOX.**

THIS BOX GIVES US A SENSE OF WHERE THE **BULK OF THE DATA SITS...**

THEN WE INDICATE THE **MINIMUM...**

...**MIDDLE...**

...AND **MAXIMUM** INDIVIDUAL VALUES WITH THESE BARS.

THAT'S ME.
265

THAT'S ME.
312

THAT'S ME.
364

Girth (in cm) 270 280 290 300 310 320 330 340 350 360 370

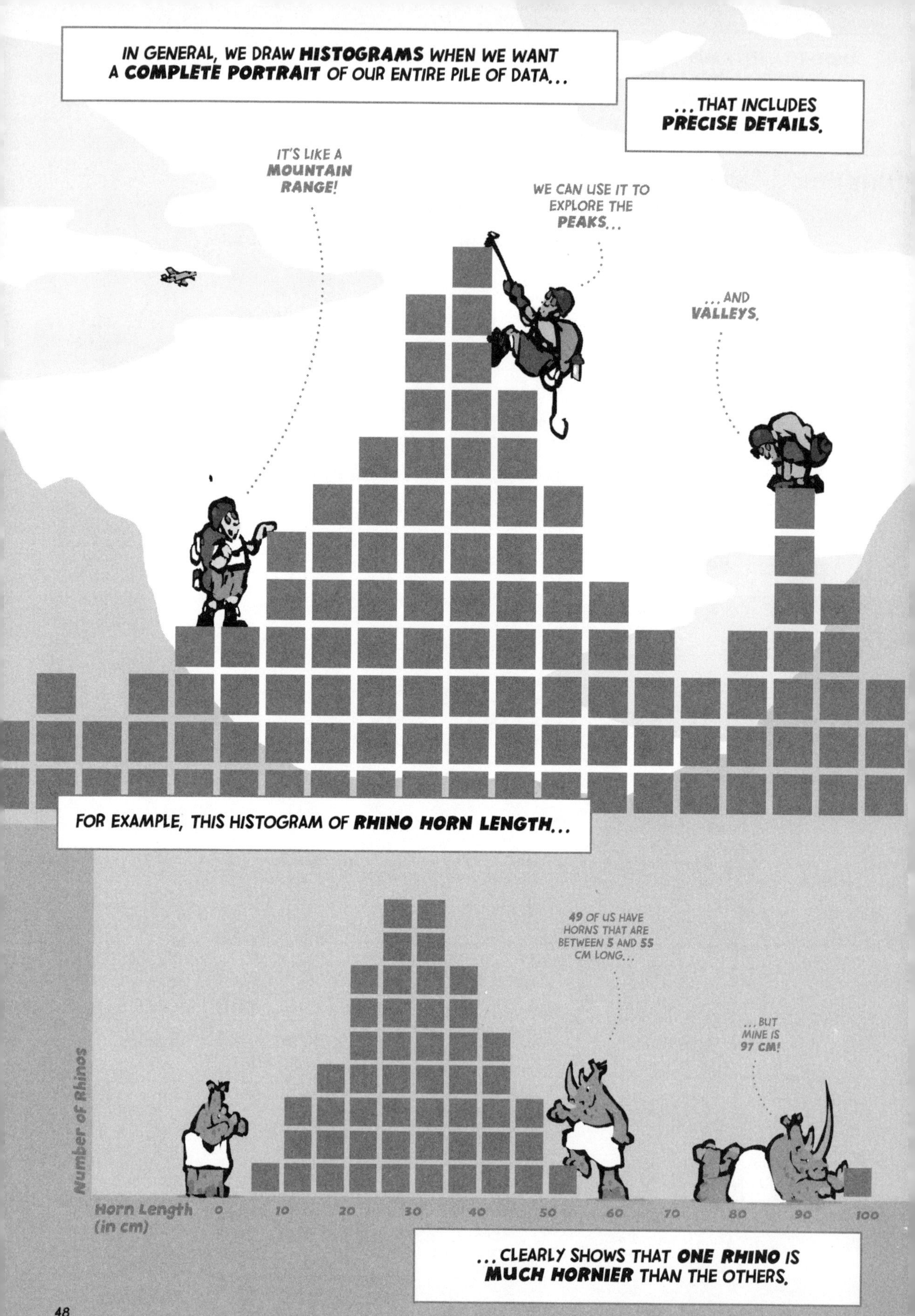

IN GENERAL, WE DRAW **HISTOGRAMS** WHEN WE WANT A **COMPLETE PORTRAIT** OF OUR ENTIRE PILE OF DATA...

...THAT INCLUDES **PRECISE DETAILS.**

IT'S LIKE A MOUNTAIN RANGE!
WE CAN USE IT TO EXPLORE THE PEAKS...
...AND VALLEYS.

FOR EXAMPLE, THIS HISTOGRAM OF **RHINO HORN LENGTH...**

49 OF US HAVE HORNS THAT ARE BETWEEN 5 AND 55 CM LONG...
...BUT MINE IS 97 CM!
Number of Rhinos
Horn Length (in cm)
0
10
20
30
40
50
60
70
80
90
100

...CLEARLY SHOWS THAT **ONE RHINO** IS **MUCH HORNIER** THAN THE OTHERS.

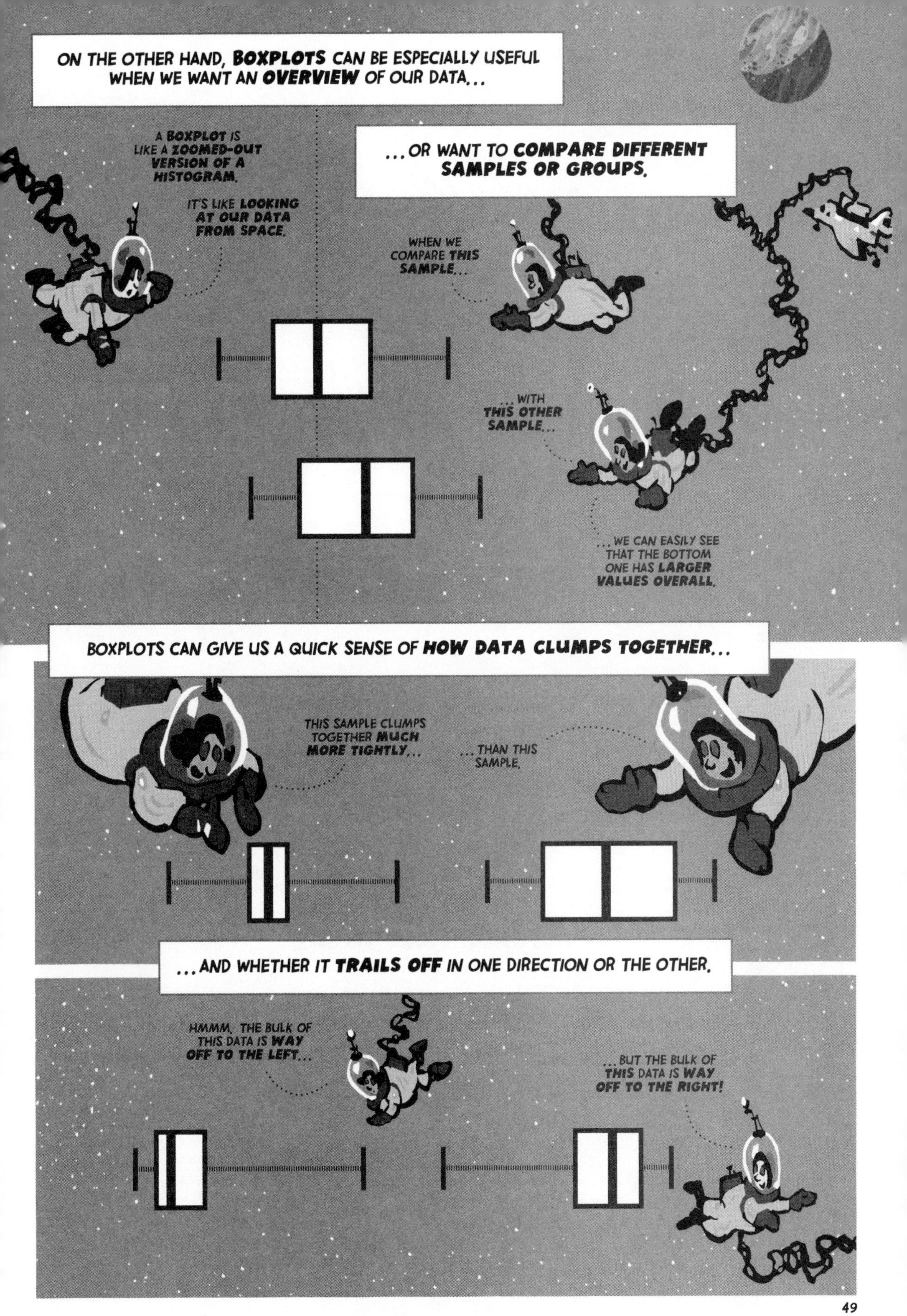

ON THE OTHER HAND, **BOXPLOTS** CAN BE ESPECIALLY USEFUL WHEN WE WANT AN **OVERVIEW** OF OUR DATA...

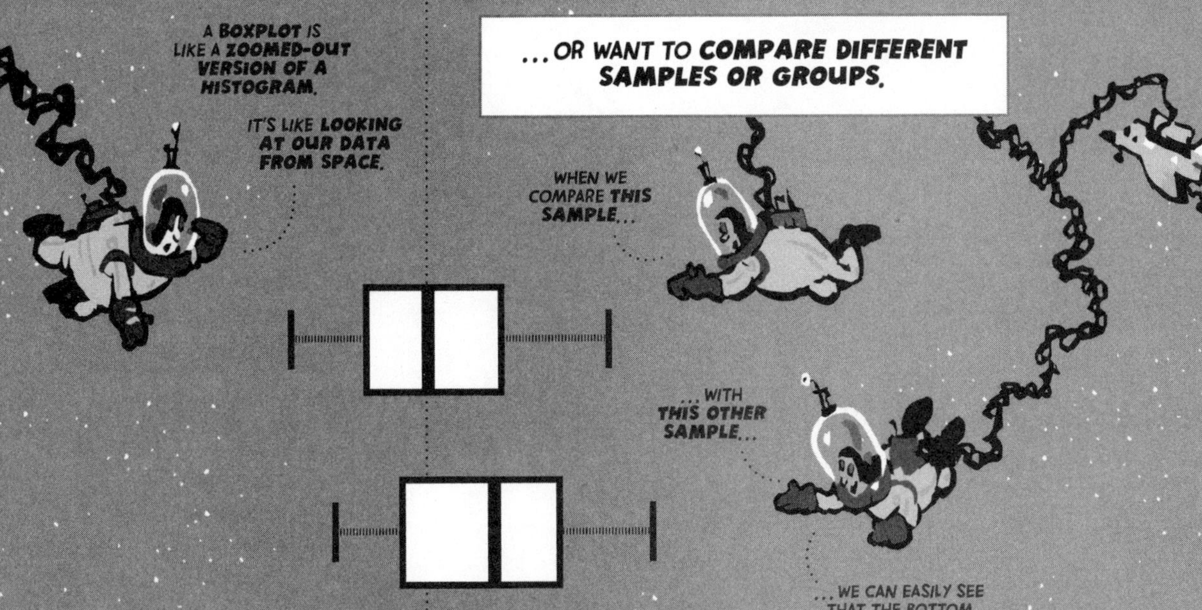

BOXPLOTS CAN GIVE US A QUICK SENSE OF **HOW DATA CLUMPS TOGETHER**...

THIS SAMPLE CLUMPS TOGETHER **MUCH MORE TIGHTLY**...

...THAN THIS SAMPLE.

...AND WHETHER IT **TRAILS OFF** IN ONE DIRECTION OR THE OTHER.

HMMM. THE BULK OF THIS DATA IS **WAY OFF TO THE LEFT**...

...BUT THE BULK OF **THIS** DATA IS **WAY OFF TO THE RIGHT!**

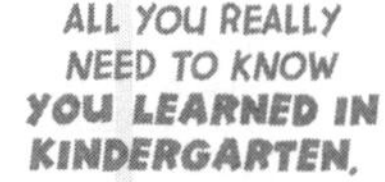

BUT THE FACT IS, THE **FIRST THING** WE SHOULD **ALWAYS** DO WITH OUR DATA IS **LOOK AT IT.**

BECAUSE ALTHOUGH WE MIGHT BE **TEMPTED** BY **MORE SOPHISTICATED MATHEMATICAL TOOLS...**

...SIMPLE PICTURES KEEP US FOCUSED ON WHAT THE DATA **ACTUALLY HAVE TO SAY.**

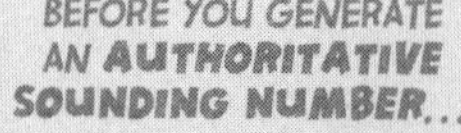

CHAPTER 4
DETECTIVE WORK

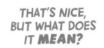

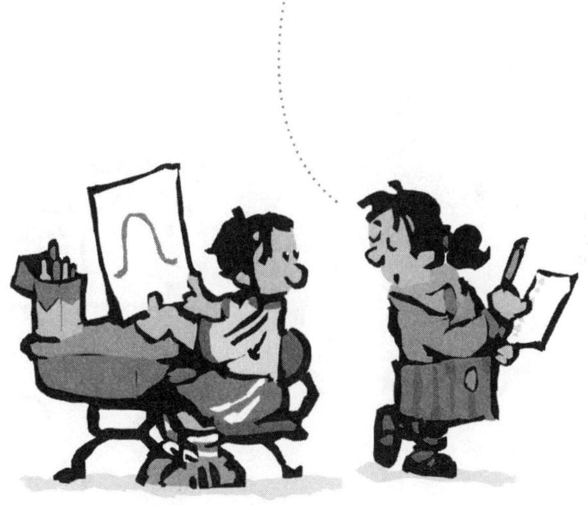
THAT'S NICE, BUT WHAT DOES IT **MEAN?**

OUR ULTIMATE GOAL IS TO **GATHER EVIDENCE** FROM **ONE RANDOM SAMPLE...**

...AND USE IT TO PIECE TOGETHER **A STORY ABOUT A POPULATION.**

GIVE ME **ONE GROUP OF RANDOM SUPERVILLAINS...**

...AND I CAN CONFIDENTLY TELL YOU SOMETHING ABOUT **ALL THE SUPERVILLAINS...**

...ALL THE SUPERVILLAINS **IN THE ENTIRE WORLD!**

BUT **FIRST** WE NEED TO LEARN HOW TO DO SOME **BASIC DETECTIVE WORK.**

WHEN WE **START TO INVESTIGATE** ANY PILE OF DATA...

THIS HISTOGRAM SHOWS 64 RANDOM SUPERVILLAINS SORTED BY **AGE**.
Supervillain Count
Age
10
20
30
40
50
60
70
80
90
100

...WE ALWAYS LOOK AT **FOUR PRIMARY CHARACTERISTICS**...

SAMPLE SIZE

HOW **MUCH DATA IS IN THERE?**

SHAPE

WHAT DOES THE PILE **LOOK LIKE?**

LOCATION

WHERE IS IT, EXACTLY?

SPREAD

HOW **WIDE** IS IT?

...AND WE'RE GOING TO SPEND THIS CHAPTER **LEARNING ABOUT THEM.**

WHAT MYSTERIES ARE HIDDEN WITHIN THIS **MOUND OF MURDERERS?**
LET'S **SIFT FOR CLUES.**

SAMPLE SIZE

HOW **MANY** **PIECES OF DATA** **ARE IN THERE?**

SAMPLE SIZE* IS THE **FIRST THING** TO LOOK FOR IN ANY PILE OF DATA...

...AND IT'S EASY TO SEE **WHY IT MATTERS.**

IF YOU HAD ONLY **A FEW** RANDOM VILLAINS IN YOUR SAMPLE...
LIKE US FIVE!
...YOU COULDN'T SAY MUCH OF **ANYTHING** ABOUT OUR OVERALL POPULATION.

SORRY, THIS PICTURE OF YOUR DATA **DOESN'T REALLY HELP MUCH.**

* SEE **PAGE 214.**

IN GENERAL, A **LARGER SAMPLE SIZE** IS **BETTER!**

AS WE'LL DISCOVER LATER, THE SIZE OF A SAMPLE IS **DIRECTLY RELATED** TO THE **LEVEL OF CONFIDENCE** WE CAN HAVE ABOUT A POPULATION.

UNFORTUNATELY, IN PRACTICE THE SIZE OF A SAMPLE IS **ALWAYS LIMITED BY SOMETHING.**

SHAPE

THE MOMENT WHEN SOMEONE DISCOVERS **THE SHAPE** OF THEIR SAMPLE CAN BE **EXTREMELY SUSPENSEFUL...**

...BECAUSE HOWEVER A PILE OF DATA IS SHAPED, IT'S **ALWAYS SHAPED THAT WAY FOR A REASON.**

FOR EXAMPLE, WE CALL A PILE OF DATA **FLAT** WHEN **ALL THE POSSIBLE OUTCOMES ARE EQUALLY LIKELY.**

WE CALL A PILE OF DATA **NORMAL** WHEN SOMETHING IS CAUSING IT TO **CLUMP AROUND ONE PARTICULAR VALUE.**

WE CALL A PILE OF DATA **SKEWED** WHEN SOMETHING IS CAUSING IT TO **TRAIL OFF MORE IN ONE DIRECTION THAN THE OTHER.**

LOCATION

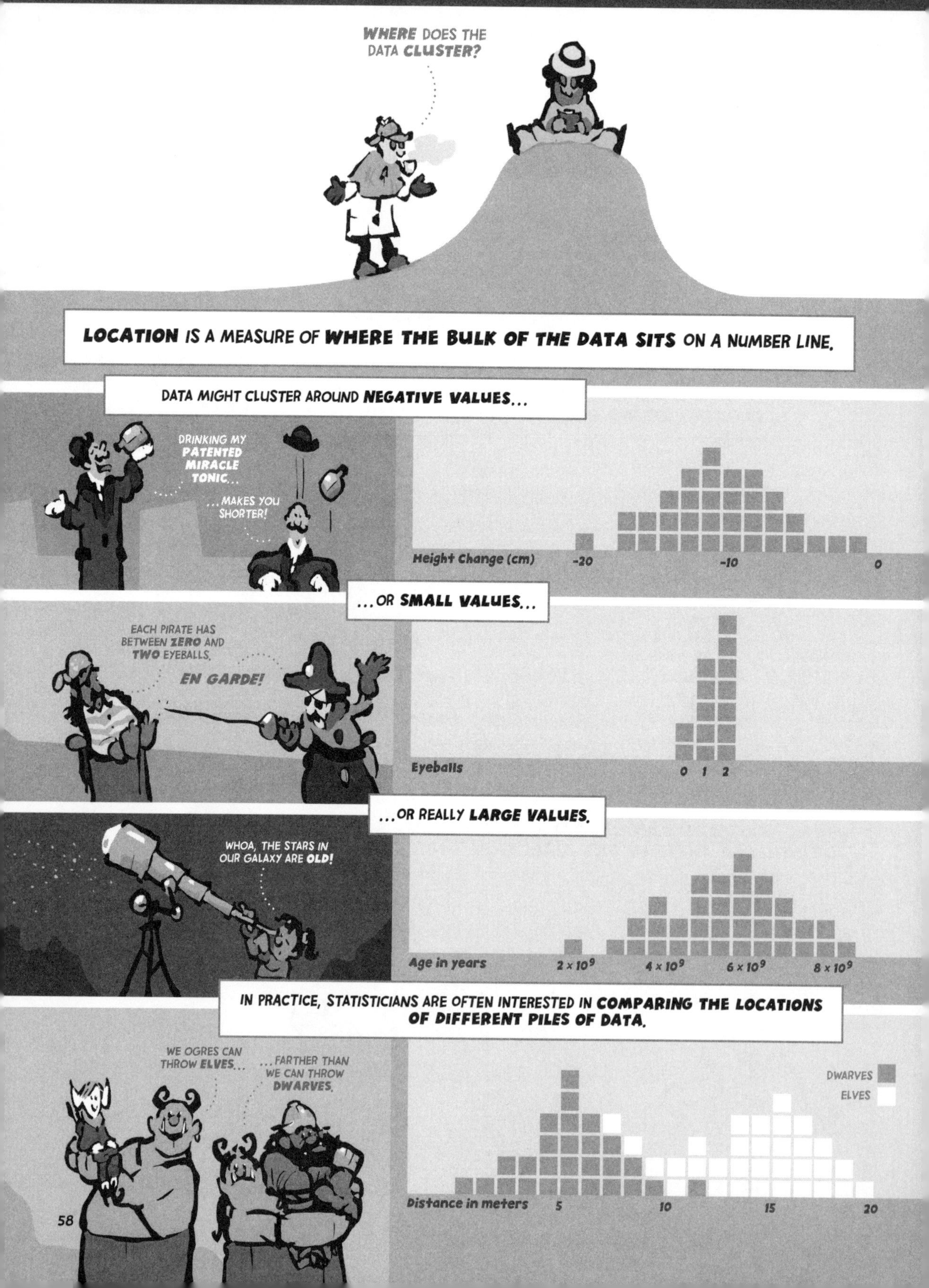

WHERE DOES THE DATA CLUSTER?

LOCATION IS A MEASURE OF **WHERE THE BULK OF THE DATA SITS** ON A NUMBER LINE.

DATA MIGHT CLUSTER AROUND **NEGATIVE VALUES...**

DRINKING MY **PATENTED MIRACLE TONIC...**

...MAKES YOU SHORTER!

Height Change (cm) -20 -10 0

...OR **SMALL VALUES...**

EACH PIRATE HAS BETWEEN **ZERO** AND **TWO** EYEBALLS.

EN GARDE!

Eyeballs 0 1 2

...OR REALLY **LARGE VALUES.**

WHOA, THE STARS IN OUR GALAXY ARE **OLD!**

Age in years 2×10^9 4×10^9 6×10^9 8×10^9

IN PRACTICE, STATISTICIANS ARE OFTEN INTERESTED IN **COMPARING THE LOCATIONS OF DIFFERENT PILES OF DATA.**

WE OGRES CAN THROW **ELVES...**

...FARTHER THAN WE CAN THROW **DWARVES.**

DWARVES

ELVES

Distance in meters 5 10 15 20

DEFINING LOCATION WITH WORDS CAN BE **TRICKY**...

...SO WE OFTEN DESCRIBE IT WITH A SINGLE NUMBER: **THE AVERAGE.***

* TECHNICALLY, THIS NUMBER IS ALSO CALLED **THE MEAN**. TO LEARN HOW TO CALCULATE IT, TURN TO **PAGE 214**.

TO CALCULATE THE AVERAGE, WE SIMPLY **ADD UP ALL THE DATA VALUES**...

...THEN **DIVIDE BY THE NUMBER OF DATA VALUES.**

HOWEVER, ALTHOUGH THE AVERAGE IS **USEFUL** AND **PRECISE** AS A MEASURE OF LOCATION, **IT ISN'T PERFECT.**

UNFORTUNATELY, **AVERAGES CAN BE DECEPTIVE.**

FOR EXAMPLE, IF A PILE OF DATA IS **SKEWED**...

...AN AVERAGE VALUE **CAN BE SERIOUSLY MISLEADING.**

WITH SKEWED DATA, THE **MEDIAN** IS OFTEN MORE REVEALING AS A MEASURE OF LOCATION...

...BECAUSE IT CAN GIVE A BETTER SENSE OF A "**TYPICAL**" **VALUE.**

SO, WHILE PEOPLE LOVE TO **CITE AVERAGES WITH AUTHORITY**...

THE AVERAGE PIRATE HAS **1.28 EYEBALLS**...
...EARNS **120 DOUBLOONS**...
...AND DRINKS **82.9** LITERS OF GROG EACH YEAR.
I'LL TAKE ALL THAT **WITH A GRAIN OF SALT**, PLEASE.

...IT'S IMPORTANT TO REMEMBER THAT AN AVERAGE TELLS US ONLY **ONE VERY PRECISE THING** ABOUT OUR DATA.

THE SUM OF ALL THE MEASUREMENTS...

WHICH IS ONE REASON WE SHOULD **NEVER** THINK ABOUT THE **LOCATION** OF ANY PILE OF DATA...

...WITHOUT ALSO THINKING ABOUT ITS **SHAPE**...

LOOK, HOLMES, THE **AVERAGE SUPERVILLAIN** IN OUR SAMPLE SCORED **510** ON THE MATH SAT!
BUT WATSON, BECAUSE **THIS BIG CLUMP** OF THEM SUCKS AT MATH...
...AND THIS **CLUMP** OF THEM ROCKS AT MATH...
...ALMOST **NONE OF THEM SCORED ANYWHERE NEAR THE AVERAGE VALUE.**
Avg
Supervillain Count
Math SAT Score
400
600
800

...AND ABOUT ITS **SPREAD**, WHICH IS COMING UP NEXT.

SPREAD

SPREAD IS A MEASURE OF THE **WIDTH OF A PILE OF DATA...**

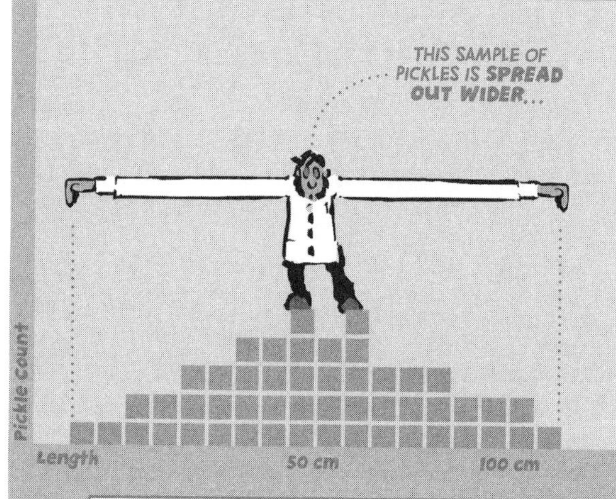

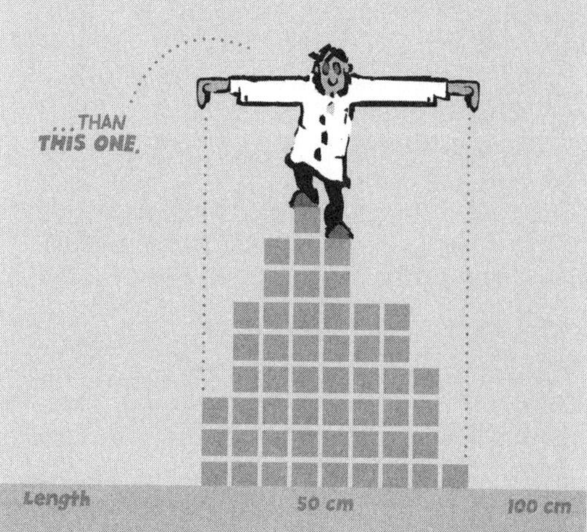

...BUT IT'S ALSO A MEASURE OF VARIATION.

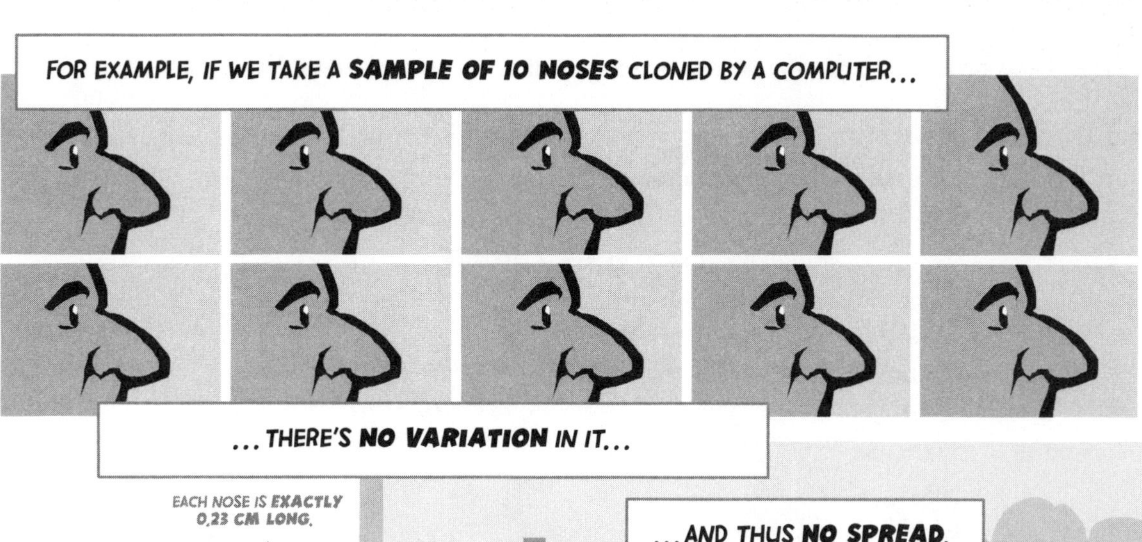
FOR EXAMPLE, IF WE TAKE A **SAMPLE OF 10 NOSES** CLONED BY A COMPUTER...

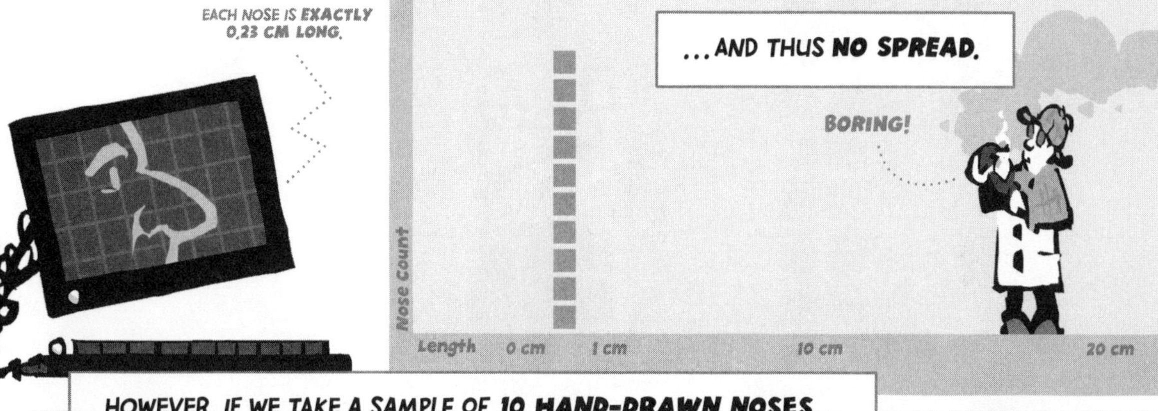
...THERE'S **NO VARIATION** IN IT...
EACH NOSE IS **EXACTLY 0.23 CM LONG.**
...AND THUS **NO SPREAD.**
BORING!
Nose Count
Length
0 cm
1 cm
10 cm
20 cm

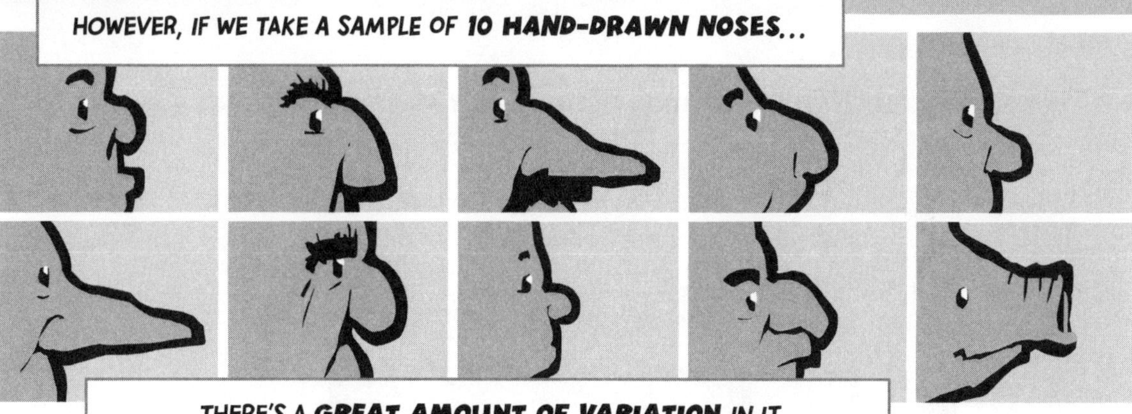
HOWEVER, IF WE TAKE A SAMPLE OF **10 HAND-DRAWN NOSES...**

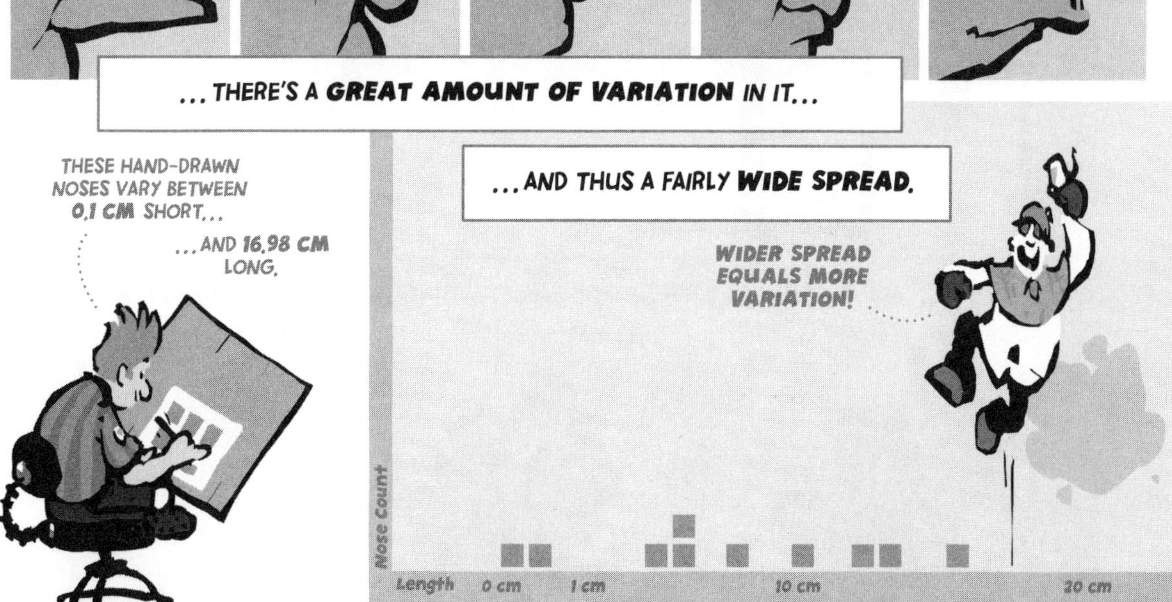
...THERE'S A **GREAT AMOUNT OF VARIATION** IN IT...
THESE HAND-DRAWN NOSES VARY BETWEEN **0.1 CM** SHORT...
...AND **16.98 CM** LONG.
...AND THUS A FAIRLY **WIDE SPREAD.**
WIDER SPREAD EQUALS MORE VARIATION!
Nose Count
Length
0 cm
1 cm
10 cm
20 cm

ONE STRAIGHTFORWARD WAY TO **MEASURE SPREAD** IS TO TAKE THE **OVERALL RANGE**...

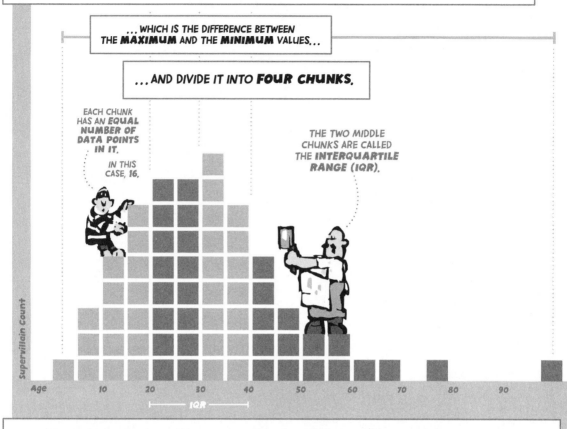

...WHICH IS THE DIFFERENCE BETWEEN THE **MAXIMUM** AND THE **MINIMUM** VALUES...

...AND DIVIDE IT INTO **FOUR CHUNKS**.

EACH CHUNK HAS AN **EQUAL NUMBER OF DATA POINTS IN IT**.

IN THIS CASE, **16**.

THE TWO MIDDLE CHUNKS ARE CALLED THE **INTERQUARTILE RANGE (IQR)**.

Supervillain Count

Age 10 20 30 40 50 60 70 80 90

IQR

THIS GIVES US A SENSE OF THE VARIATION **WITHIN EACH PART** OF THE OVERALL SAMPLE...

BELOW THE MEDIAN, THINGS ARE PRETTY **COMPRESSED**...
...ABOVE THE MEDIAN, THINGS ARE **MORE SPREAD OUT**.

...AND IS ESPECIALLY USEFUL FOR INVESTIGATING DATA THAT ARE **SKEWED**.

LOOKS LIKE OUR SUPERVILLAINS **SKEW OLD**, WATSON...
...BUT I WONDER HOW THINGS WILL LOOK IF **JIMMY THE GEEZER** EVER KICKS THE BUCKET?
I'M AN **OUTLIER**.

THE MOST COMMON MEASURE OF SPREAD, HOWEVER, IS STANDARD DEVIATION (SD).*
IF WE THINK OF THE AVERAGE AS A CENTRAL VALUE...
...THE STANDARD DEVIATION IS BASED ON AN AVERAGE DISTANCE FROM THAT VALUE.
Avg
SD
SD WORKS BEST WHEN THE PILE OF DATA IS FAIRLY SYMMETRICAL.
LIKE THIS ONE, WHICH DESCRIBES OUR HEIGHT.
STANDARD MEANS TYPICAL...
...DEVIATION MEANS DIFFERENCE!
Supervillain Count
Height (in cm)
140
160
180
200
UNFORTUNATELY, CALCULATING STANDARD DEVIATION IS A BIT TRICKY.
WE TAKE THE SQUARE ROOT OF THE AVERAGE SQUARED DIFFERENCE FROM THE AVERAGE VALUE!
AAHHHH!
SO FOR NOW JUST REMEMBER THAT A WIDER PILE OF DATA HAS A LARGER STANDARD DEVIATION.
AND A LARGER STANDARD DEVIATION...
...MEANS MORE VARIATION!
Avg
SD

* SEE **PAGE 215** TO LEARN HOW TO CALCULATE IT.

IN THIS CHAPTER WE LEARNED ABOUT THE **FOUR IMPORTANT QUALITIES** TO LOOK FOR IN ANY **ONE SAMPLE...**

...AND SOON WE'LL BE HUNTING FOR SOME OF **THESE SAME QUALITIES** IN A LARGER **POPULATION.**

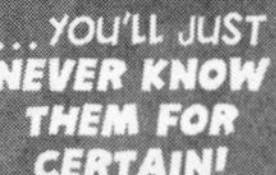

BUT FIRST, LET'S USE WHAT WE'VE LEARNED SO FAR TO **RESOLVE A CONTROVERSY!**

CHAPTER 5
MONSTER MISTAKES

MOST OF THE TIME WHEN WE GO OUT AND GATHER DATA...

...IT'S BECAUSE WE'RE INVESTIGATING AN IMPORTANT QUESTION ABOUT THE WORLD.

WHEN DID THESE MOUNTAINS RISE UP FROM THE SEA?

HOW MANY BEHEADINGS HAPPENED DURING THE REIGN OF KING HENRY VIII?

WILL GIRLS DIG IT IF I WEAR THESE PANTS?

SOME QUESTIONS ARE FAIRLY **STRAIGHTFORWARD**...

...AND CAN BE TACKLED BY LOOKING AT ONLY **ONE SET OF SAMPLE DATA.**

HOW MANY PEOPLE IN THIS COUNTRY HAVE DIABETES?
LET'S EXAMINE 100 RANDOM CITIZENS AND MAKE A GUESS.
DO VAMPIRES HAVE BAD BREATH?
LET'S EXAMINE 100 RANDOM VAMPIRES AND MAKE A GUESS.

BUT OTHER QUESTIONS ARE MORE **COMPARATIVE**...

WHEN THEY BITE DIABETIC PEOPLE...
...DO VAMPIRES GET BAD BREATH?

...AND REQUIRE MORE **COMPLEX ANALYSIS.**

MORE **COMPLEX** STATISTICS PROBLEMS OFTEN INVOLVE **EXPLORING A RELATIONSHIP...**

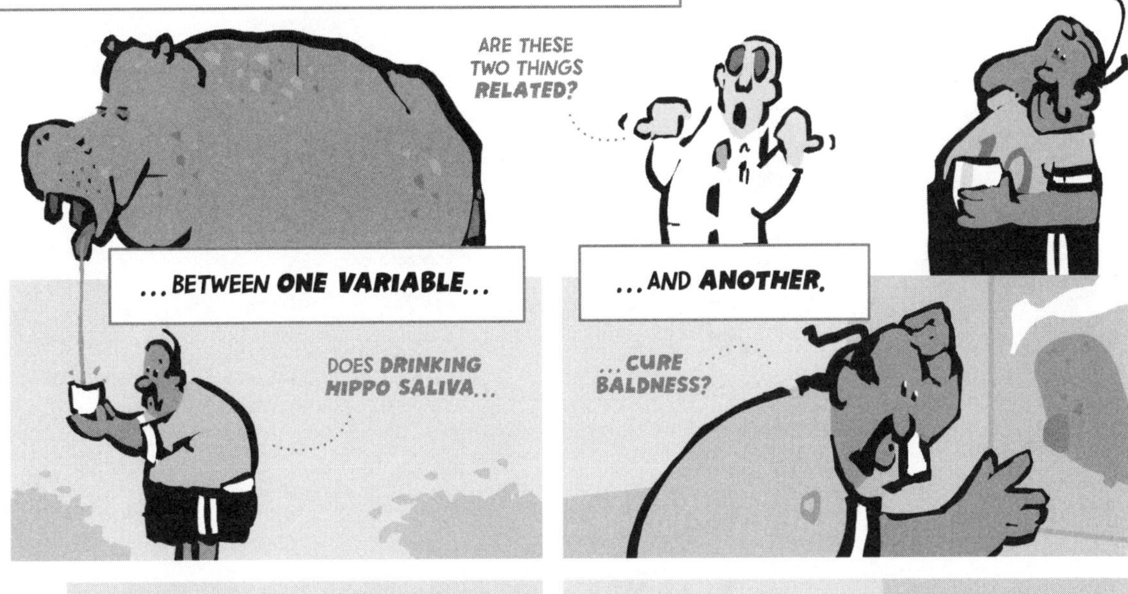

MUCH OF THE TIME WE'RE EXPLORING HOW ONE VARIABLE **INFLUENCES** ANOTHER...

...BUT REMEMBER, WHEN WE USE STATISTICS WE CAN NEVER TOTALLY PROVE **ANY** OF OUR CONCLUSIONS.

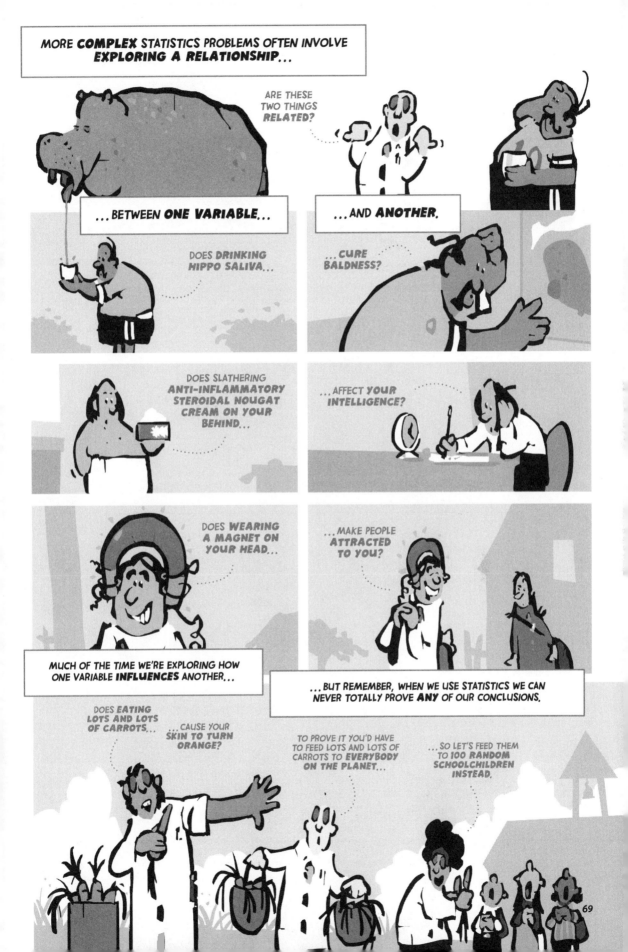

BUT RECENTLY, **FEMALE VIKINGS** HAVE STARTED RIDING...

...AND THEY'RE CONVINCED THAT **THEY'RE FASTER.**

* SEE **PAGE 215** FOR A TECHNICAL DEFINITION.

...HAS A NOTICEABLE EFFECT ON **SPEED**...

...VIKING JUDGES **COLLECTED SOME DATA.**

THEY **RANDOMLY SELECTED 50 MALE** RIDERS...

...AND **50 FEMALE** RIDERS...

...AND TIMED RIDERS AS THEY ZIPPED AROUND **A KILOMETER-LONG COURSE.**

ALL TIMES ARE IN **SECONDS.**

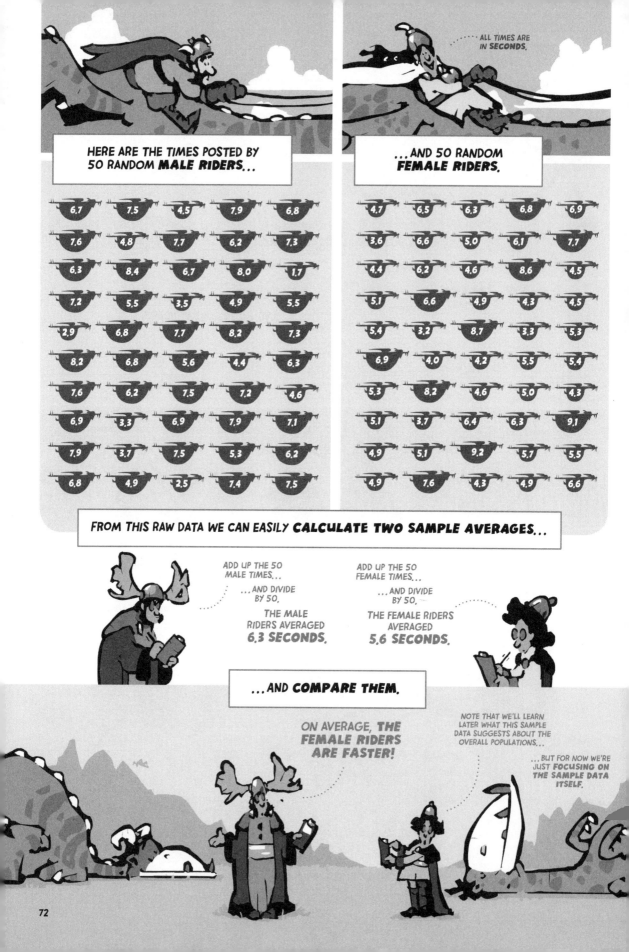

HERE ARE THE TIMES POSTED BY 50 RANDOM **MALE RIDERS...**

...AND 50 RANDOM **FEMALE RIDERS.**

FROM THIS RAW DATA WE CAN EASILY **CALCULATE TWO SAMPLE AVERAGES...**

ADD UP THE 50 MALE TIMES...
...AND DIVIDE BY 50.
THE MALE RIDERS AVERAGED **6.3 SECONDS.**

ADD UP THE 50 FEMALE TIMES...
...AND DIVIDE BY 50.
THE FEMALE RIDERS AVERAGED **5.6 SECONDS.**

...AND **COMPARE THEM.**

THIS COMPARISON **MIGHT** SEEM **TO SETTLE THE MATTER.**

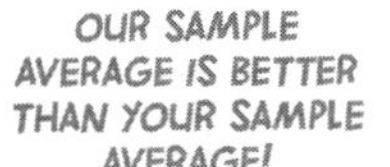

OUR SAMPLE AVERAGE IS BETTER THAN YOUR SAMPLE AVERAGE!

BUT SO FAR WE'VE LOOKED AT ONLY **ONE PART OF THE OVERALL PICTURE.**

BEWARE THE **HASTY AVERAGE!**
WE ALWAYS NEED TO LOOK AT SHAPE, LOCATION, **AND** SPREAD.

TO GET A MORE DETAILED SENSE OF THE DATA, WE SHOULD **ALWAYS DRAW PICTURES.**

GATHER ROUND AND LET'S **SEE WHAT THE NUMBERS REVEAL.**

SURE ENOUGH, A DIFFERENT PICTURE EMERGES...

BOTH GROUPS ARE SKEWED!

...WHEN WE COMPARE A BOXPLOT OF THE MALE RIDERS' DATA...
WE HAD THE SLOWER TIMES OVERALL...
...BUT WE HAD MORE VARIATION OVER THERE ON THE FASTER SIDE!
MEDIAN
Seconds
1
2
3
4
5
6
7
8
9
10

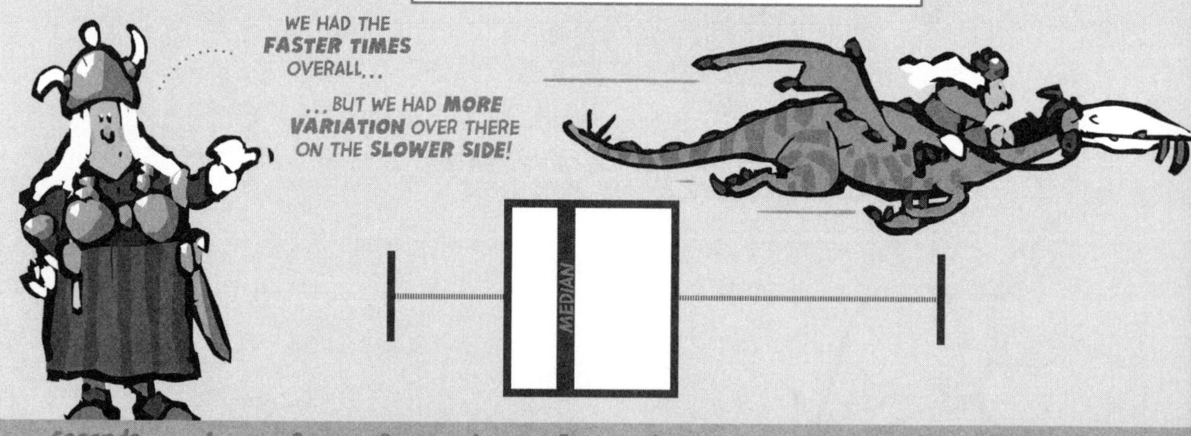
...WITH A BOXPLOT OF THE FEMALE RIDERS' DATA.
WE HAD THE FASTER TIMES OVERALL...
...BUT WE HAD MORE VARIATION OVER THERE ON THE SLOWER SIDE!
MEDIAN
Seconds
1
2
3
4
5
6
7
8
9
10

WHY WOULD BOTH GROUPS BE SKEWED...
...IN DIFFERENT DIRECTIONS?

I SUSPECT HIDDEN FORCES AT WORK!

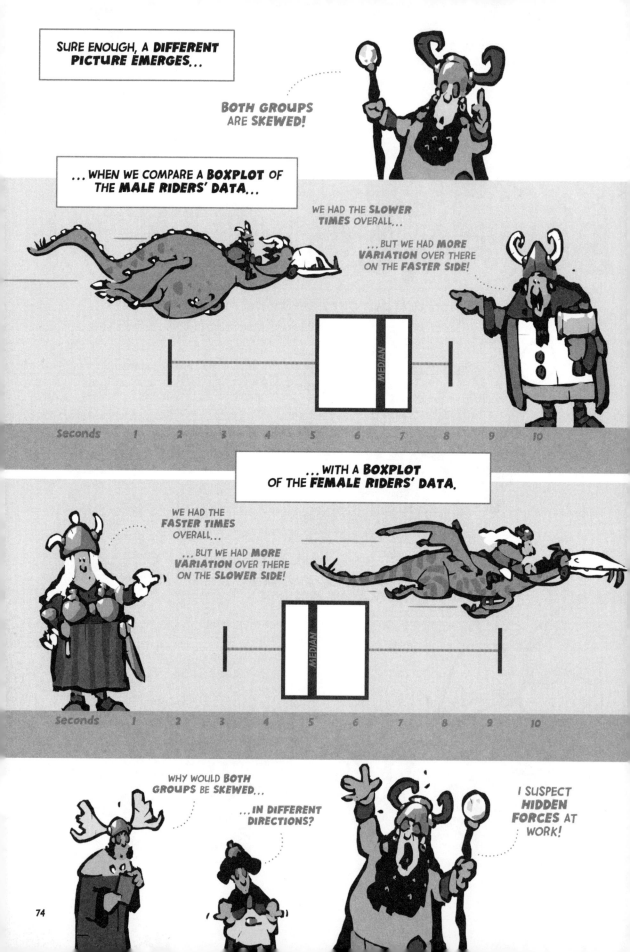

AND THE **MYSTERY DEEPENS...**

...WHEN WE LOOK AT **HISTOGRAMS** OF THE **MALE DATA...**

...AND THE **FEMALE DATA.**

THESE PICTURES SUGGEST THAT THE RELATIONSHIP BETWEEN OUR TWO VARIABLES **MAY NOT BE AS SIMPLE AS WE THOUGHT.**

IF **BEING A WOMAN...**

...MAKES YOU **GO FASTER...**

...WHY DO **BOTH** DATA SETS HAVE **SKEWED TAILS AND MYSTERIOUS DOUBLE HUMPS?**

THE CHALLENGE NOW IS TO FIGURE OUT **WHY THE DATA LOOKS THE WAY IT DOES...**

SKEWED TAILS AND DOUBLE HUMPS?
I THINK WE'RE MISSING SOMETHING MONSTROUS.

...BY SEARCHING FOR **OTHER VARIABLES THAT MIGHT BE INFLUENCING IT.**

WHAT ELSE COULD BE AFFECTING RIDER SPEED?

COULD IT BE SOMETHING ABOUT THE COURSE?
COULD BE, BUT I DOUBT IT...
...SINCE THAT WAS THE SAME FOR BOTH SETS OF RIDERS.

COULD IT BE HOW MUCH THE RIDERS WEIGH...
...OR WHAT THEY WEAR?
COULD BE, BUT I DOUBT IT.
REMEMBER, WE CHOSE THEM RANDOMLY.

IT TURNS OUT THAT WHILE FOCUSING ON **GENDER** AND **SPEED...**

...WE'VE BEEN NEGLECTING TO THINK ABOUT **THE DRAGONS THEMSELVES!**

WE'RE A THIRD VARIABLE!

THE FACT IS, THERE ARE **TWO SEPARATE SPECIES OF DRAGON...**

...AND MALE RIDERS TEND TO PREFER THE LARGER, **SLOWER** DRAGONS...

...WHILE FEMALE RIDERS PREFER THE SMALLER, **FASTER** DRAGONS!

SO IT'S NO WONDER THE FEMALE RIDERS **SEEMED FASTER OVERALL.**

WHEN WE ACCOUNT FOR **DRAGON PREFERENCE...**

...BY CALCULATING AVERAGE TIMES BY **BOTH RIDER GENDER AND DRAGON TYPE...**

...THE RESULTS ARE **SURPRISING.**

	SMALL DRAGONS	BIG DRAGONS
MALE RIDERS	3.6	6.9
FEMALE RIDERS	5.1	7.9

IT TURNS OUT THAT OUR **FIRST CONCLUSION...**

...WAS NOT ONLY **MISLEADING...**

...IT WAS **DEAD WRONG!**

...A **HIDDEN, LURKING VARIABLE** WAS SNEAKING AROUND...

UNFORTUNATELY, LURKING VARIABLES CAN **BEDEVIL ALL KINDS OF STATISTICAL ANALYSIS**...

THE **MORAL** OF THIS STORY...

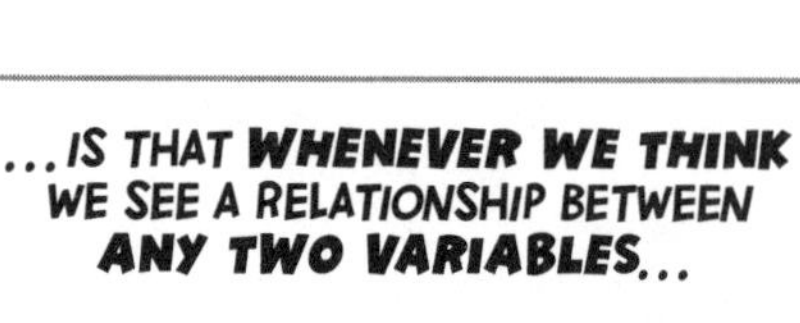

...THERE MIGHT BE SOME OTHER VARIABLE **INFLUENCING OUR CONCLUSIONS...**

...AND IF WE DON'T FIND IT, WE RISK BELIEVING THINGS THAT **AREN'T TRUE!**

CHAPTER 6
FROM SAMPLES TO POPULATIONS

THUS FAR, WE'VE BEEN TALKING ALMOST EXCLUSIVELY ABOUT **SAMPLES.**

BUT REMEMBER, OUR **ULTIMATE TASK** IS TO USE **SAMPLES...**

...TO MAKE **CONFIDENT STATEMENTS** ABOUT **ENTIRE POPULATIONS.**

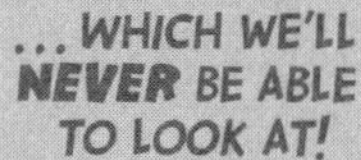

THIS POSES A **PROBLEM:**

HOW CAN WE BE CONFIDENT ABOUT POPULATIONS...
...WHEN WE'LL **NEVER** BE ABLE TO LOOK AT THEM?
IT'S **MURKY** DOWN THERE.

IN **PART TWO OF THIS BOOK** WE'RE GOING TO TACKLE THIS PROBLEM HEAD-ON...

WE'RE GOING TO LEARN **STATISTICAL INFERENCE!**

...BUT BEFORE WE BEGIN, LET'S CLARIFY SOME OF THE **KEY TERMS** WE'LL BE USING.

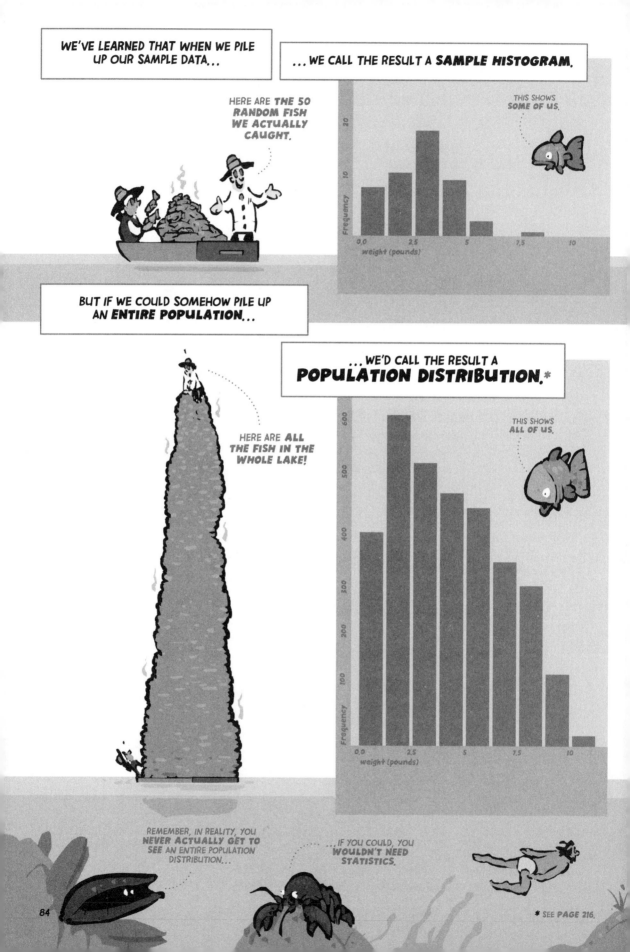

WE'VE LEARNED THAT WHEN WE PILE UP OUR SAMPLE DATA...

...WE CALL THE RESULT A **SAMPLE HISTOGRAM.**

HERE ARE **THE 50 RANDOM FISH WE ACTUALLY CAUGHT.**

THIS SHOWS **SOME OF US.**

Frequency: 10, 20

weight (pounds): 0.0, 2.5, 5, 7.5, 10

BUT IF WE COULD SOMEHOW PILE UP AN **ENTIRE POPULATION...**

...WE'D CALL THE RESULT A **POPULATION DISTRIBUTION.***

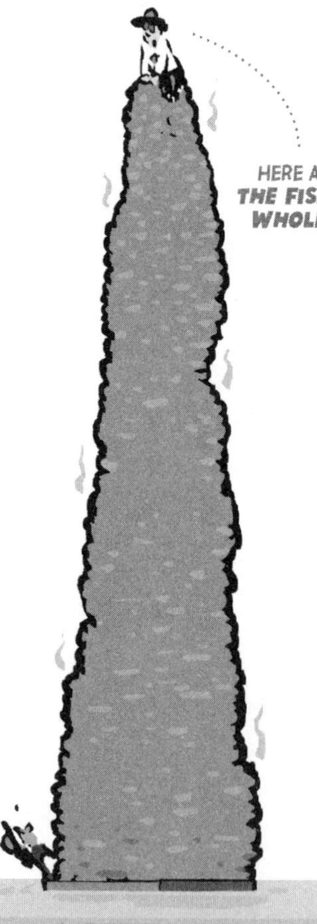

HERE ARE **ALL THE FISH IN THE WHOLE LAKE!**

THIS SHOWS **ALL OF US.**

Frequency: 100, 200, 300, 400, 500, 600

weight (pounds): 0.0, 2.5, 5, 7.5, 10

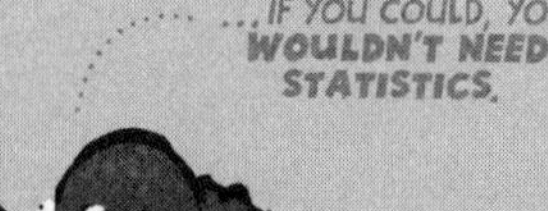

REMEMBER, IN REALITY, YOU **NEVER ACTUALLY GET TO SEE** AN ENTIRE POPULATION DISTRIBUTION...
...IF YOU COULD, YOU **WOULDN'T NEED STATISTICS.**

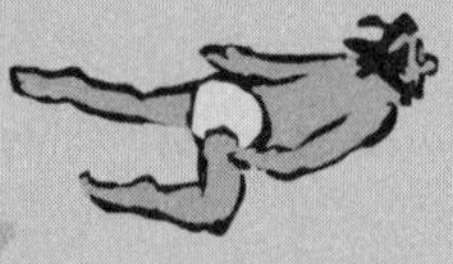

* SEE **PAGE 216.**

SIMILARLY, WE'VE LEARNED THAT **SAMPLE HISTOGRAMS HAVE CERTAIN IMPORTANT QUALITIES...**

THE PILE OF FISH WE CAUGHT...
...HAS A SHAPE...
...LOCATION...
...AND SPREAD...
AVG =3.7
SD=1.9
...AND WE KNOW THEM!

...AND IT TURNS OUT THAT **POPULATION DISTRIBUTIONS ALSO HAVE THESE QUALITIES.**

THE ENTIRE POPULATION OF FISH IN THE LAKE...
?
...ALSO HAS A SHAPE, LOCATION, AND SPREAD...
AVG =?
SD=?
...BUT WE'LL NEVER KNOW THEM FOR CERTAIN.

IN ORDER TO DISTINGUISH BETWEEN THEM, WE REFER TO QUALITIES IN **SAMPLES** AS **"STATISTICS"...**

FOR EXAMPLE, OUR **SAMPLE AVERAGE IS A STATISTIC...**

...AND SO IS OUR **SAMPLE STANDARD DEVIATION.**

...AND TO QUALITIES IN **POPULATIONS** AS **"PARAMETERS."***

FOR EXAMPLE, OUR **OVERALL POPULATION AVERAGE IS A PARAMETER...**

...AND SO IS OUR **POPULATION STANDARD DEVIATION.**

* SEE **PAGE 216.**

IN OTHER WORDS, THE ONLY REASON WE GO OUT AND **GATHER STATISTICS**...

...IS BECAUSE WE'RE **CURIOUS ABOUT PARAMETERS.**

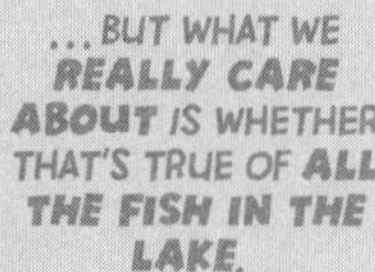

STATISTICS ARE THE THINGS WE **ACTUALLY CALCULATE AND THEREFORE KNOW WITH CERTAINTY**...

...AND **PARAMETERS** ARE THE THINGS WE REALLY **WANT TO KNOW** BUT CAN **ONLY MAKE GUESSES ABOUT.**

LUCKILY, ALTHOUGH WE'LL **NEVER** BE ABLE TO **LOOK AT PARAMETERS DIRECTLY**...

...WE CAN **USE STATISTICS** TO **HUNT FOR THEM.**

YOU NAME IT!
BUT WE'RE GOING TO FOCUS ON **ONE IN PARTICULAR.**

WE'RE GOING TO LEARN TO USE STATISTICS WE FIND IN **ONE RANDOM SAMPLE...**

...TO HUNT FOR THE **AVERAGE IN THE POPULATION IT COMES FROM.**

SAMPLE SIZE?
CHECK.
SAMPLE AVERAGE?
CHECK.
SAMPLE STANDARD DEVIATION?
CHECK.
OKAY, WE'RE READY TO HUNT!
IT'S GOT TO BE AROUND HERE SOMEWHERE!

AS WE KNOW, WE'LL **NEVER** BE ABLE TO USE **STATISTICS...**
...TO ACHIEVE **CERTAINTY ABOUT PARAMETERS.**
I CAN EXAMINE THESE **50 RANDOM ALIENS** I JUST GATHERED FROM THAT GALAXY.
BUT THERE ARE A **GAZILLION MORE** ALIENS IN THIS GALAXY THAT SHE'LL NEVER BE ABLE TO EXAMINE.
FORTUNATELY, STATISTICIANS HAVE DISCOVERED A WAY TO **BRIDGE THE DISTANCE BETWEEN THEM...**
WHOA! WE CAN USE **THIS SHAPE** LIKE A **LENS...**
...TO **BRING POPULATIONS INTO FOCUS!**
...AND WE'RE GOING TO SPEND THE **NEXT CHAPTER LEARNING ABOUT IT!**

PART TWO
HUNTING PARAMETERS

CHAPTER 7
THE CENTRAL LIMIT THEOREM

THIS CHAPTER IS ABOUT THE **GREAT DISCOVERY**...
...THAT MAKES **EVERYTHING IN THE REST OF THIS BOOK POSSIBLE**...
...AND IT HAS TO DO WITH **AVERAGES**.

LET'S IMAGINE THAT WE WANT TO KNOW THE **AVERAGE VALUE IN A CERTAIN POPULATION.**

HOW MUCH **SODA** DOES EACH AMERICAN **DRINK PER DAY?**

SLURP.

BURP!

THEN LET'S IMAGINE WE GO OUT AND GATHER **A WHOLE BUNCH OF SEPARATE RANDOM SAMPLES FROM THAT POPULATION.**

HERE ARE **50 RANDOM AMERICANS.**

HERE ARE 50 **OTHER** RANDOM AMERICANS.

HERE ARE 50 **OTHER** RANDOM AMERICANS.

EACH SAMPLE HAS **50 RANDOM AMERICANS** IN IT.

WE PUT **EACH SAMPLE IN A BAG** TO HELP US KEEP TRACK OF THEM.

OUCH, **STOP POKING!**

IT TURNS OUT THAT IF WE CALCULATE THE **AVERAGE VALUE IN EACH OF OUR RANDOM SAMPLES**...
FOR EXAMPLE, THE AVERAGE IN OUR SAMPLE IS 17.2 OUNCES.
IN OUR SAMPLE IT'S 12.9 OUNCES.
HERE, IT'S 18.4 OUNCES.
6.3
17.2
12.9
18.4
6.3
15.3
16.1

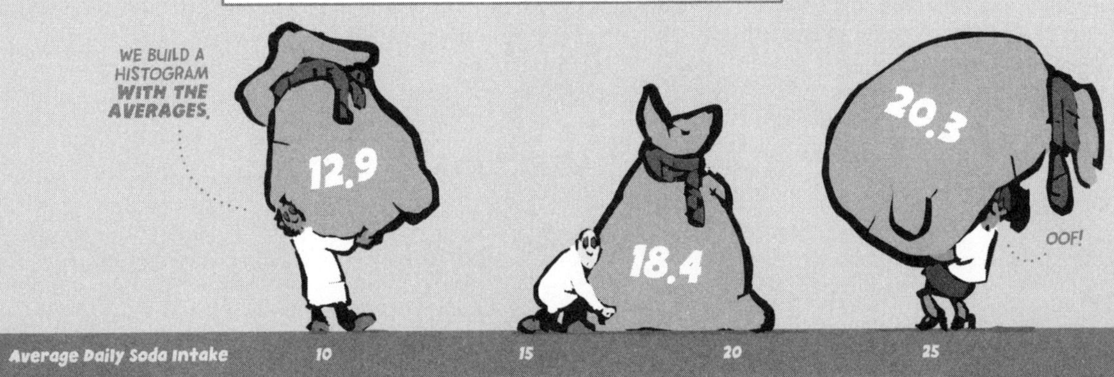

...THEN ORDER THEM AND **PILE THEM UP**...
WE BUILD A HISTOGRAM **WITH THE AVERAGES.**
12.9
18.4
20.3
OOF!
Average Daily Soda Intake
10
15
20
25

...**THE PILE OF AVERAGES** WILL EVENTUALLY START TO **CLUMP TOGETHER!**

AND THAT'S **NOT ALL.**

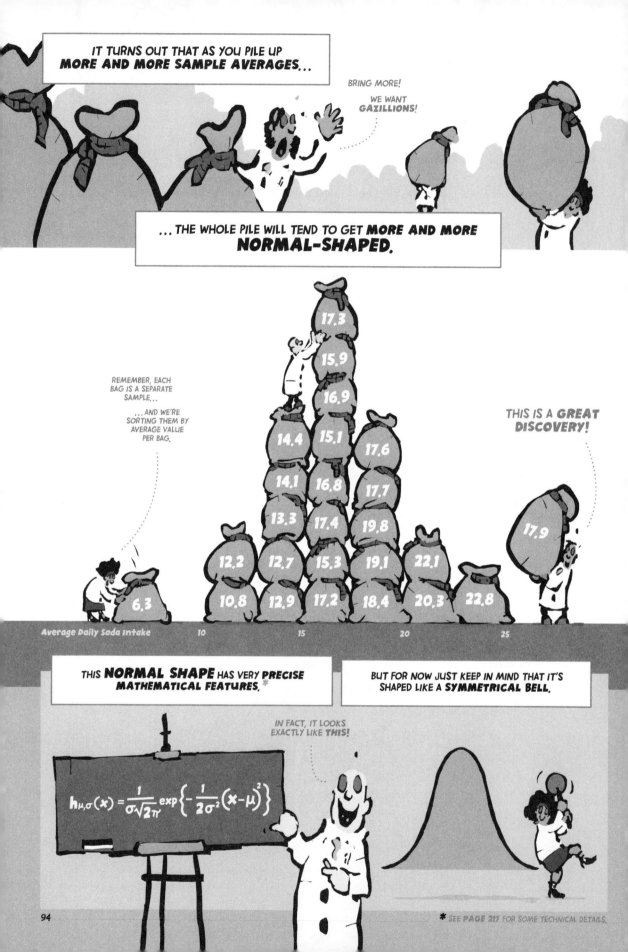

IT TURNS OUT THAT AS YOU PILE UP **MORE AND MORE SAMPLE AVERAGES...**

BRING MORE!
WE WANT **GAZILLIONS!**

...THE WHOLE PILE WILL TEND TO GET **MORE AND MORE NORMAL-SHAPED.**

REMEMBER, EACH BAG IS A SEPARATE SAMPLE...
...AND WE'RE SORTING THEM BY AVERAGE VALUE PER BAG.
THIS IS A **GREAT DISCOVERY!**
17.3
15.9
16.9
14.4
15.1
17.6
14.1
16.8
17.7
13.3
17.4
19.8
17.9
12.2
12.7
15.3
19.1
22.1
6.3
10.8
12.9
17.2
18.4
20.3
22.8
Average Daily Soda Intake
10
15
20
25

THIS **NORMAL SHAPE** HAS VERY **PRECISE MATHEMATICAL FEATURES.***

BUT FOR NOW JUST KEEP IN MIND THAT IT'S SHAPED LIKE A **SYMMETRICAL BELL.**

IN FACT, IT LOOKS EXACTLY LIKE **THIS!**
$h_{\mu,\sigma}(x) = \frac{1}{\sigma\sqrt{2\pi}} \exp\left\{-\frac{1}{2\sigma^2}(x-\mu)^2\right\}$

* SEE **PAGE 217** FOR SOME TECHNICAL DETAILS.

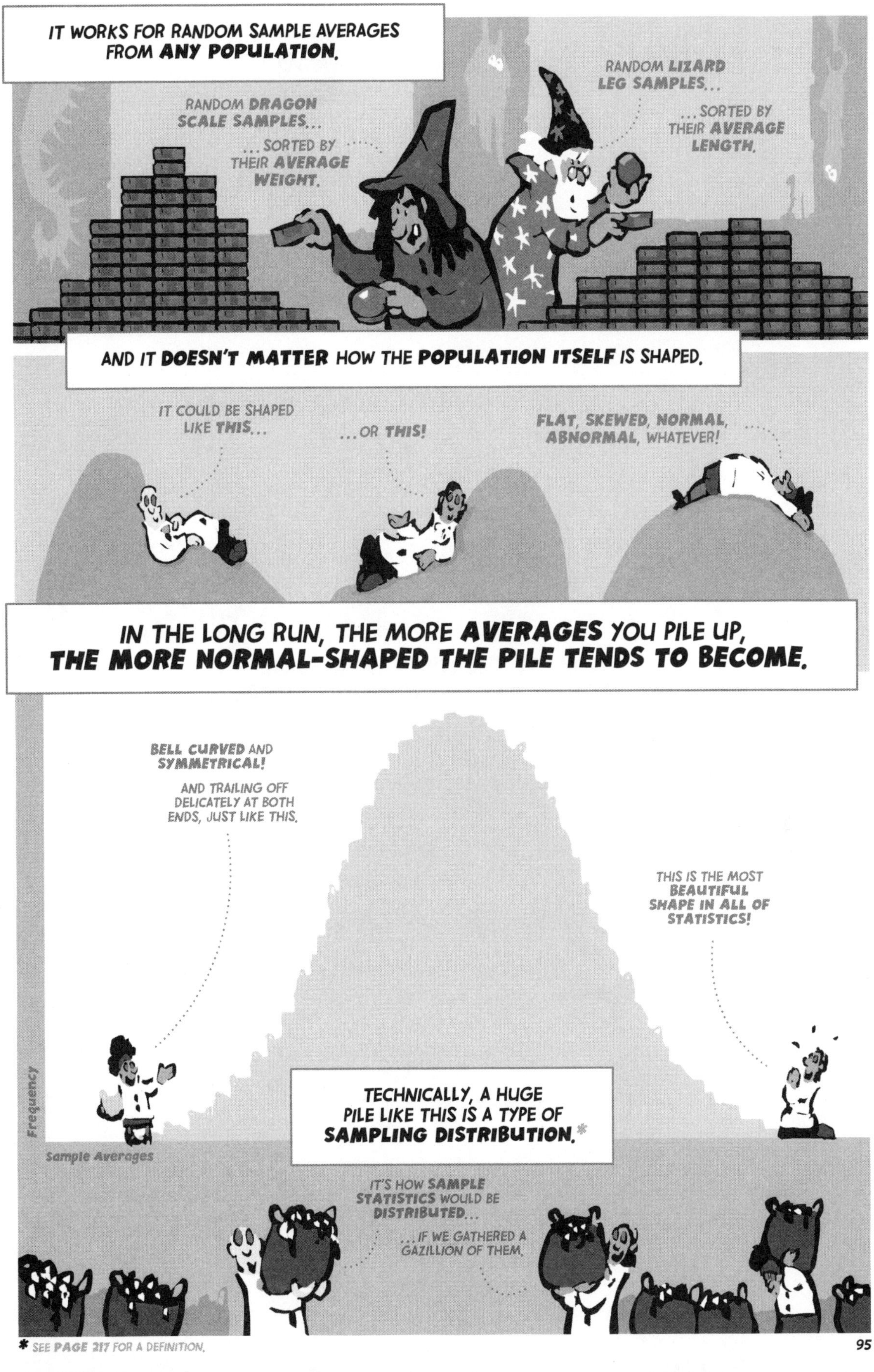

IT WORKS FOR RANDOM SAMPLE AVERAGES FROM **ANY POPULATION.**

RANDOM **DRAGON SCALE SAMPLES...**

...SORTED BY THEIR **AVERAGE WEIGHT.**

RANDOM **LIZARD LEG SAMPLES...**

...SORTED BY THEIR **AVERAGE LENGTH.**

AND IT **DOESN'T MATTER** HOW THE **POPULATION ITSELF** IS SHAPED.

IT COULD BE SHAPED LIKE **THIS...**

...OR **THIS!**

FLAT, SKEWED, NORMAL, ABNORMAL, WHATEVER!

IN THE LONG RUN, THE MORE **AVERAGES** YOU PILE UP, **THE MORE NORMAL-SHAPED THE PILE TENDS TO BECOME.**

BELL CURVED AND **SYMMETRICAL!**

AND TRAILING OFF DELICATELY AT BOTH ENDS, JUST LIKE THIS.

THIS IS THE MOST **BEAUTIFUL SHAPE IN ALL OF STATISTICS!**

Frequency

Sample Averages

TECHNICALLY, A HUGE PILE LIKE THIS IS A TYPE OF **SAMPLING DISTRIBUTION.***

IT'S HOW **SAMPLE STATISTICS** WOULD BE **DISTRIBUTED...**

...IF WE GATHERED A GAZILLION OF THEM.

* SEE **PAGE 217** FOR A DEFINITION.

PLUS, THERE'S **AN ADDED BONUS:**

IT TURNS OUT THAT THE **CENTER VALUE** IN A **GIANT PILE OF AVERAGES...**

...IS EQUAL TO THE **CENTER VALUE OF THE POPULATION IT CAME FROM.**

THE AVERAGE OF ALL THE AVERAGES...

...EQUALS **THE POPULATION AVERAGE.**

THE POPULATION MIGHT BE SHAPED LIKE **THIS...**

...WE'LL **NEVER KNOW FOR CERTAIN.**

FOR EXAMPLE, IF THIS GIANT PILE OF SAMPLE AVERAGES SORTED BY DAILY SODA INTAKE IS CENTERED AT 17 **OUNCES PER DAY**...

...THE OVERALL POPULATION WILL BE **CENTERED AT THAT SAME VALUE!**

ADSI
16
17
18
19

ADSI 15 16 17 18 19

THIS WORKS BECAUSE A GIANT PILE OF AVERAGES IS **GUARANTEED TO BE SYMMETRICAL.**

NORMAL DISTRIBUTIONS **ARE ALWAYS SYMMETRICAL.**

IN THE LONG RUN, FOR **EACH** SAMPLE AVERAGE WE GRAB WITH A VALUE **BELOW THE POPULATION AVERAGE**...

...WE'RE GUARANTEED TO EVENTUALLY GRAB **ANOTHER** SAMPLE AVERAGE WITH A VALUE **ABOVE THE POPULATION AVERAGE.**

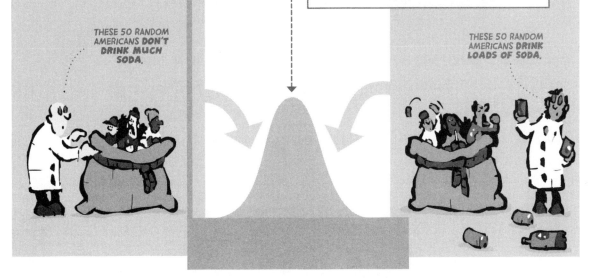

THESE 50 RANDOM AMERICANS DON'T DRINK MUCH SODA.

THESE 50 RANDOM AMERICANS DRINK LOADS OF SODA.

AND THERE'S A **SECOND ADDED BONUS:**

IT TURNS OUT THAT **A GIANT PILE OF AVERAGES...**

SD SD

REMEMBER, THIS PILE HAS A **GAZILLION** SAMPLES IN IT.

...WILL ALSO TEND TO BE NARROWER THAN THE POPULATION IT CAME FROM.

SD SD

EXACTLY **HOW MUCH NARROWER** DEPENDS ON **THE SIZE OF EACH SAMPLE.**

AS WE INCREASE THE SAMPLE SIZE, THE GIANT PILE WILL LOOK LESS LIKE **THIS...**

...AND MORE LIKE **THIS.**

NOTE THAT **BOTH OF THESE PILES ARE NORMAL-SHAPED.**

BUT THE SLENDER ONE HAS A **SMALLER STANDARD DEVIATION.**

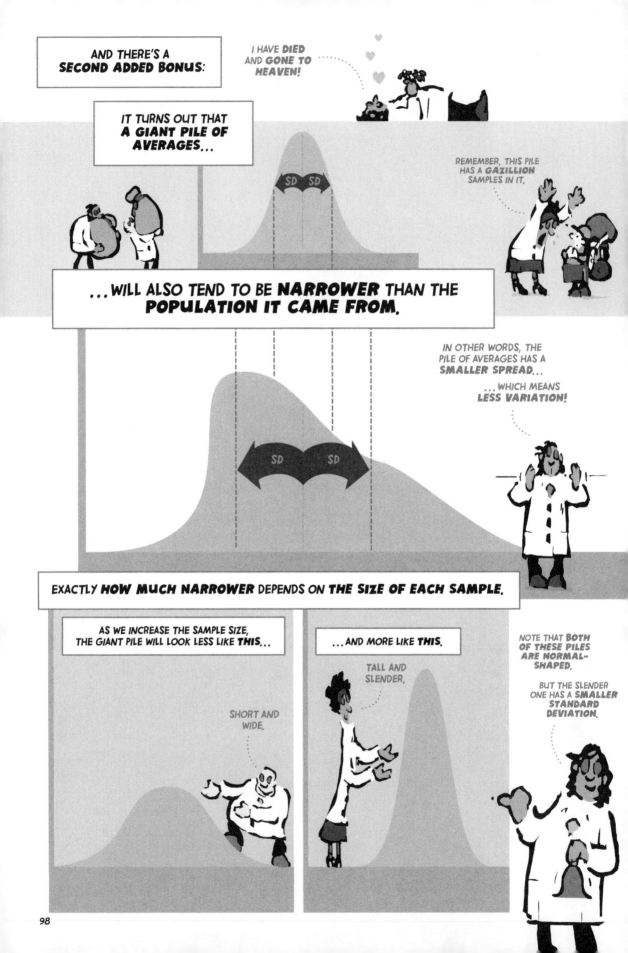

THERE'S AN INTUITIVE WAY TO THINK ABOUT WHY A **LARGER SAMPLE SIZE** RESULTS IN A **NARROWER PILE OF AVERAGES.**

IF EACH SAMPLE HAS ONLY **ONE AMERICAN** IN IT...

...THEN THE SPREAD OF THE PILE OF AVERAGES **WILL BE EXACTLY THE SAME AS THE SPREAD OF THE WHOLE POPULATION.**

THE VARIATION BETWEEN BAGS...

...WILL EQUAL THE VARIATION BETWEEN INDIVIDUALS IN THE OVERALL POPULATION!

BUT IF EACH SAMPLE HAS **ALL AMERICANS** IN THE WHOLE POPULATION JAMMED INTO IT...

...THEN THE SPREAD OF THE PILE OF AVERAGES WILL BE **ZERO.**

THERE WILL BE NO VARIATION BETWEEN BAGS!
DUH!

IN ANY CASE, THE MATHEMATICAL RELATIONSHIP IS **VERY PRECISE.**

THE STANDARD DEVIATION OF THE ENORMOUS PILE...
...EQUALS THE POPULATION STANDARD DEVIATION...
...DIVIDED BY THE SQUARE ROOT OF THE SAMPLE SIZE!

AAHHHH!

HERE'S A SUMMARY OF THE **GREAT DISCOVERY:**

IN THE LONG RUN, **GIANT PILES OF RANDOM SAMPLE AVERAGES TEND TO BE NORMAL-SHAPED!**

THEY'RE **CENTERED AROUND THE POPULATION AVERAGE...**

...BUT **NARROWER THAN THE POPULATION.**

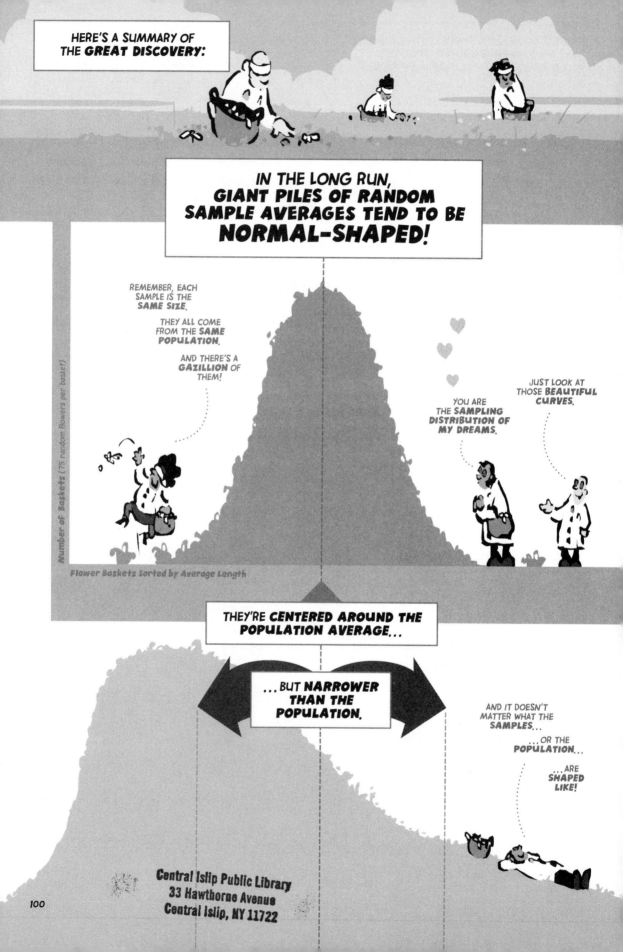

TECHNICALLY, WE CALL THIS DISCOVERY **THE CENTRAL LIMIT THEOREM (CLT).***

* SEE **PAGES 217–218** TO READ ABOUT THE CLT AND THESE CONDITIONS IN GREATER DETAIL.

I WISH IT HAD A **MORE POETIC NAME.**

OVER THE YEARS, STATISTICIANS HAVE THROWN TOGETHER SOME MATH THAT PROVES EXACTLY **WHY THE CLT WORKS.**

DOUBLE, DOUBLE TOIL AND TROUBLE;
RANDOM AVERAGE SAMPLES BUBBLE.
EYE OF NEWT!
TOE OF FROG!
SCALE OF DRAGON!
TONGUE OF DOG!

BUT THEY'VE ALSO DISCOVERED THAT THERE ARE **A FEW CONDITIONS.**

IT **ONLY WORKS** IF EACH SAMPLE IS TAKEN **RANDOMLY...**

THERE CAN BE NOTHING **EXCEPT CHANCE** THAT MAKES ANY ONE SAMPLE DIFFERENT FROM ANY OTHER SAMPLE.

...AND IF EACH SAMPLE IS **LARGE ENOUGH.**

A SAMPLE SIZE OF **30 OR MORE** IS USUALLY SUFFICIENT...

...BUT IT DEPENDS ON SOME OTHER COMPLICATED MATH.

* CHECK OUT **PAGE 217** TO LEARN HOW TO SAY ALL THIS WITH MATH SYMBOLS.

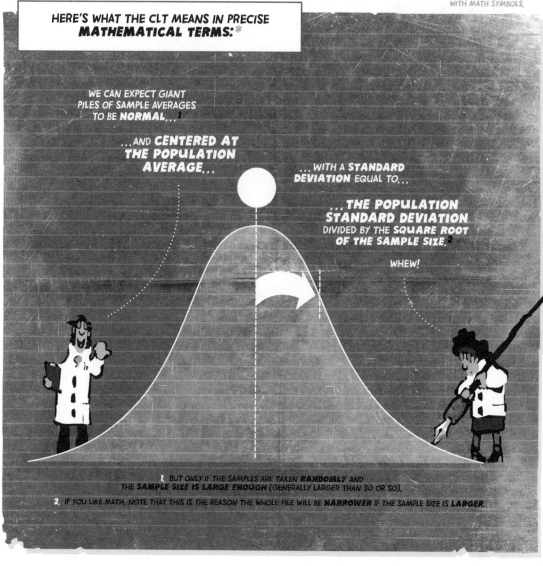

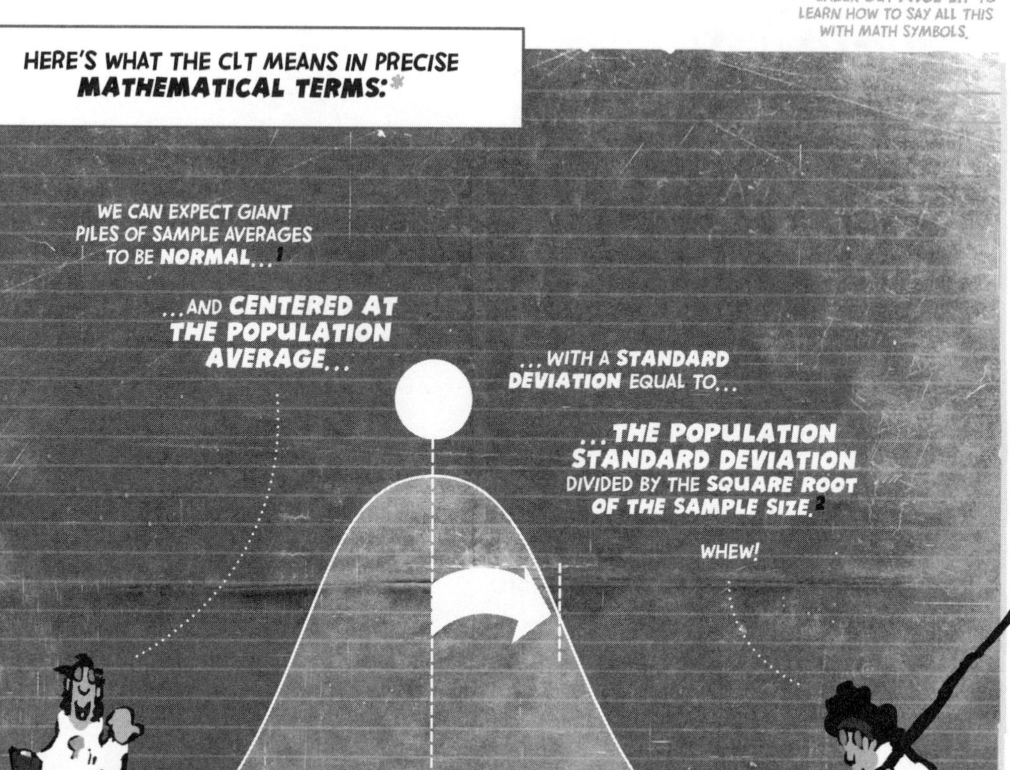
HERE'S WHAT THE CLT MEANS IN PRECISE **MATHEMATICAL TERMS:***
WE CAN EXPECT GIANT PILES OF SAMPLE AVERAGES TO BE **NORMAL**...[1]
...AND **CENTERED AT THE POPULATION AVERAGE**...
...WITH A **STANDARD DEVIATION** EQUAL TO...
...**THE POPULATION STANDARD DEVIATION** DIVIDED BY THE **SQUARE ROOT OF THE SAMPLE SIZE**.[2]
WHEW!
1. BUT ONLY IF THE SAMPLES ARE TAKEN **RANDOMLY** AND THE **SAMPLE SIZE IS LARGE ENOUGH** (GENERALLY LARGER THAN 30 OR SO).
2. IF YOU LIKE MATH, NOTE THAT THIS IS THE REASON THE WHOLE PILE WILL BE **NARROWER** IF THE SAMPLE SIZE IS **LARGER**.

KEEP THIS BLUEPRINT, BECAUSE WE'LL BE **USING IT LATER**.

BUT HERE'S A **SIMPLER WAY TO REMEMBER IT:**

IN THE NEXT SEVERAL CHAPTERS, WE'RE GOING TO **LEARN WHY IT MATTERS.**

IT GIVES US SOMETHING WE CAN BE CONFIDENT ABOUT!

WE DON'T KNOW **EVERYTHING**...
...BUT THAT DOESN'T MEAN WE **KNOW NOTHING!**

CHAPTER 8
PROBABILITIES

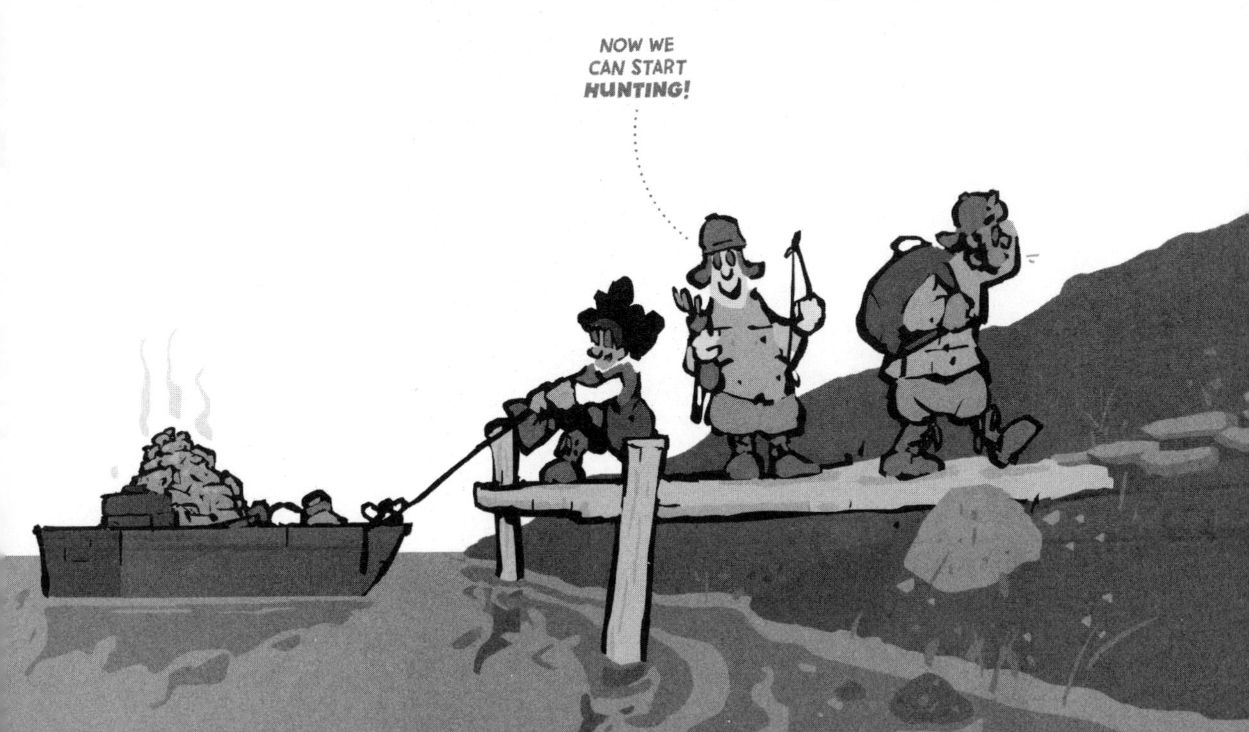
NOW WE CAN START **HUNTING!**

IN THE LAST CHAPTER, WE LEARNED THAT **GIANT PILES OF SAMPLE AVERAGES...**

...TEND TO HAVE A **NORMAL SHAPE.**
Number of Jars (500 random olives per jar)
Olive Jars sorted by average weight per jar
WE CAN BE **CONFIDENT** ABOUT THIS...
...IF THE SAMPLES ARE TAKEN **RANDOMLY...**
...**AND** THE SAMPLE SIZE IS **LARGE ENOUGH!**
NOW WE'RE GOING TO LEARN **WHY THAT MATTERS...**
WHAT MAKES **THAT SHAPE** SO **SPECIAL?**

...BY EXAMINING A **GIANT PILE OF SAMPLE AVERAGES...**

...INSIDE **CRAZY BILLY'S BAIT BARN.**

Crazy Billy's BAIT BARN

HI, I'M **BILLY.**

CAREFUL...

...HE'S **CRAZY!**

CRAZY BILLY IS CALLED CRAZY BILLY BECAUSE HE SPENDS AN **INSANE** AMOUNT OF TIME **CATCHING RANDOM WORM SAMPLES...**

30 WORMS PER SAMPLE.
I GRAB THEM **TOTALLY RANDOMLY...**
...FROM THE OVERALL POPULATION IN THE SWAMP.

...PUTTING EACH SAMPLE INTO A CAN...

BEFORE I SEAL EACH CAN I MEASURE THE WORMS...

...AND **CALCULATE THE AVERAGE WORM LENGTH PER CAN.**

...AND VERY CAREFULLY **PILING UP GAZILLIONS OF THOSE CANS,** EACH ACCORDING TO ITS AVERAGE VALUE...

...INSIDE **HIS ENORMOUS BAIT BARN...**

THIS CAN HAS AN AVERAGE LENGTH OF **4.75 INCHES,** SO I'LL PUT IT **EXACTLY HERE.**

Number of Cans

Average Length per Can (inches) 3.5 4 4.5 5

...OR **SO HE CLAIMS.**

YOU HAVE AN **ACTUAL SAMPLING DISTRIBUTION** IN THERE?

YUP, IT'S ALL BEHIND THAT **DOOR.**

IN THIS CHAPTER WE'RE GOING TO FIND OUT **WHAT WE CAN LEARN FROM CRAZY BILLY'S HUGE PILE.**

MORE SPECIFICALLY, WE'RE GOING TO FOCUS ON **THIS QUESTION:**

IF WE HAVE ACCESS **ONLY** TO **WHAT'S INSIDE THE BARN...**

...WHAT CAN WE SAY ABOUT THE **OVERALL POPULATION OF WORMS IN THE SWAMP?**

HERE'S THE SAME QUESTION IN **TECHNICAL TERMS:**

THE FIRST **COOL THING** WE COULD FIGURE OUT IF WE COULD PEEK INTO BILLY'S BARN...

...IS THE OVERALL **POPULATION AVERAGE.**

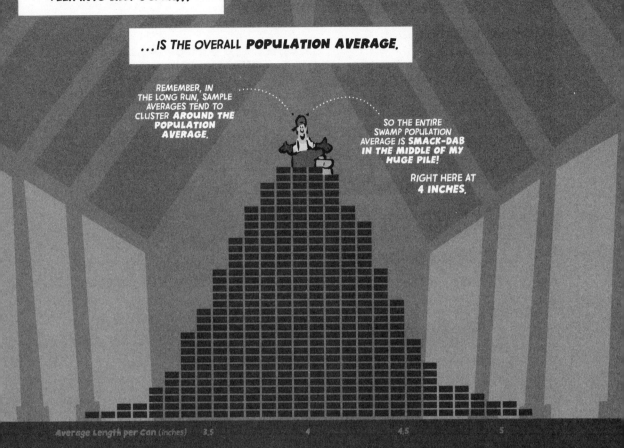

REMEMBER, IN THE LONG RUN, SAMPLE AVERAGES TEND TO CLUSTER **AROUND THE POPULATION AVERAGE.**

SO THE ENTIRE SWAMP POPULATION AVERAGE IS **SMACK-DAB IN THE MIDDLE OF MY HUGE PILE!**

RIGHT HERE AT **4 INCHES.**

Average Length per Can (inches) 3.5 4 4.5 5

IN OTHER WORDS, IF WE WERE OUT **HUNTING FOR THE POPULATION AVERAGE IN THE SWAMP...**

...WE COULD **LOOK INSIDE THE BARN** TO FIND IT!

WHAT'S THE **AVERAGE LENGTH OF THE WORMS IN THIS SWAMP?**

NO NEED TO **SOIL YOUR SUIT** DIGGING IN THE MUCK.

BUT **THAT'S NOT ALL...**

THE **OTHER COOL THING** WE COULD DO IF WE COULD PEEK INTO BILLY'S BARN...

AND THIS IS THE **REALLY IMPORTANT** COOL THING!

...IS **CALCULATE PROBABILITIES ABOUT THE OVERALL POPULATION!**

Crazy Billy's
BAIT BARN
WHAT'S A **PROBABILITY?**
IT'S A FANCY WORD FOR **LIKELIHOOD** OR **CHANCE.**

HERE'S HOW IT WORKS:

IF WE WERE ABLE TO **COUNT ALL THE CANS IN BILLY'S HUGE PILE...**

...AND FIND THAT **50% OF THEM** HAVE AN AVERAGE VALUE **WITHIN THIS RANGE...**

REMEMBER, EACH CAN HAS **30 RANDOM WORMS** IN IT.

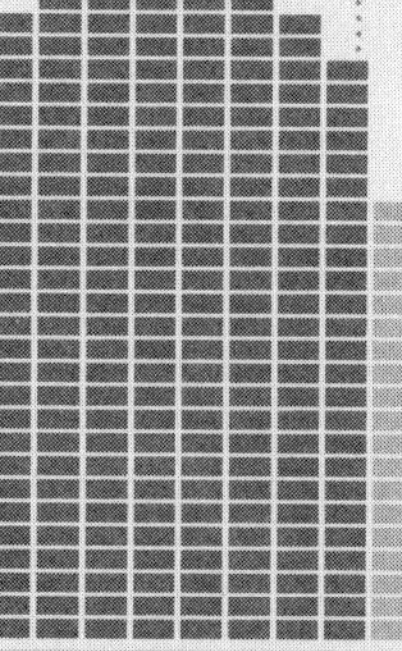

THAT'S ALL THE CANS IN THIS **DARKER SHADED PORTION...**
...WITH AVERAGES BETWEEN **3.75 AND 4.25 INCHES LONG.**

...IT WOULD MEAN THAT IF WE **RANDOMLY ASSEMBLED ONE CAN FROM THE POPULATION...**

...THERE WOULD BE A **50% PROBABILITY** IT WOULD HAVE AN AVERAGE VALUE **WITHIN THAT SAME RANGE!**

30 RANDOM WORMS, COMING RIGHT UP!
WHICH MEANS THERE'S A **50% CHANCE** IT'LL HAVE AN AVERAGE WORM LENGTH BETWEEN **3.75 AND 4.25 INCHES LONG!**

ALTERNATIVELY, IF WE COUNTED ALL THE CANS AND DISCOVERED **THIS ABOUT BILLY'S PILE:**

IT WOULD MEAN **THIS ABOUT THE OVERALL POPULATION:**

AND IF **THIS** WERE TRUE ABOUT THE PILE:

THIS WOULD BE TRUE ABOUT THE POPULATION:

IN OTHER WORDS, BY **PEEKING INSIDE THE BARN...**

...WE CAN **MEASURE OUR CHANCES** OF RANDOMLY GATHERING RANGES OF AVERAGES **FROM THE SWAMP!**

THERE ARE A FEW THINGS TO KEEP IN MIND WHEN WE **CALCULATE PROBABILITIES.***

* SEE **PAGE 218** FOR MORE TECHNICAL DETAILS.

FIRST, PROBABILITIES APPLY ONLY TO THE **LONG RUN**...

...SO THEY'LL **NEVER GIVE US CERTAINTY ABOUT THE SHORT RUN.**

SECOND, EVERY PROBABILITY HAS A **FLIP SIDE**...

...BECAUSE PROBABILITIES ALWAYS **ADD UP TO 100%.**

IF THERE'S A **95% CHANCE** ONE THING WILL HAPPEN...

...THERE'S ALWAYS GONNA BE A **5% CHANCE SOMETHING ELSE** WILL HAPPEN.

FINALLY, BY DEFINITION, WE CAN CALCULATE PROBABILITIES ONLY **ABOUT RANDOM EVENTS...**

BY DEFINITION, A **PROBABILITY** IS A NUMBER THAT **QUANTIFIES THE LONG-TERM LIKELIHOOD** THAT A **CERTAIN RANDOM EVENT** WILL OCCUR.

...AND THAT'S WHY WE ALWAYS **GATHER STATISTICS RANDOMLY.**

IF I DIDN'T GATHER MY WORMS **RANDOMLY...**
...THE PILE IN MY BARN WOULDN'T MEAN DIDDLY.

MORE GENERALLY, WE CAN CALCULATE PROBABILITIES ABOUT **OTHER RANDOM EVENTS,** LIKE **COIN FLIPS...**

THE PROBABILITY OF **FLIPPING A COIN** AND GETTING **HEADS...**

...BECAUSE **IN THE LONG RUN,** WE CAN EXPECT **50% OF ALL COIN FLIPS TO LAND ON HEADS.**

...AND **ROLLS OF THE DICE.**

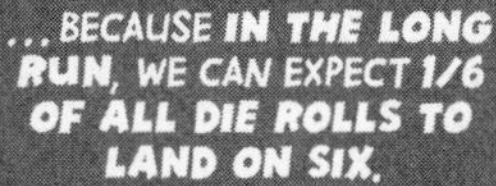

THE PROBABILITY OF **ROLLING A DIE** AND GETTING **A SIX...**

...BECAUSE **IN THE LONG RUN,** WE CAN EXPECT **1/6 OF ALL DIE ROLLS TO LAND ON SIX.**

BUT LET'S RETURN TO **RANDOM WORM HUNTING...**

BLINDFOLDS ON!

...BECAUSE WE HAVE **ONE MORE REALLY IMPORTANT THING TO LEARN ABOUT BILLY'S BARN!**

IT TURNS OUT THAT WE DON'T HAVE TO ACTUALLY **COUNT** ALL THE INDIVIDUAL CANS IN BILLY'S BARN...

... TO CALCULATE PROBABILITIES.

IT TURNS OUT THAT BECAUSE WE **KNOW** BILLY'S PILE IS **NORMAL-SHAPED**...

... WE CAN USE SOME **FANCY MATH** TO FIGURE OUT EXACTLY HOW THE CANS FIT INSIDE IT!

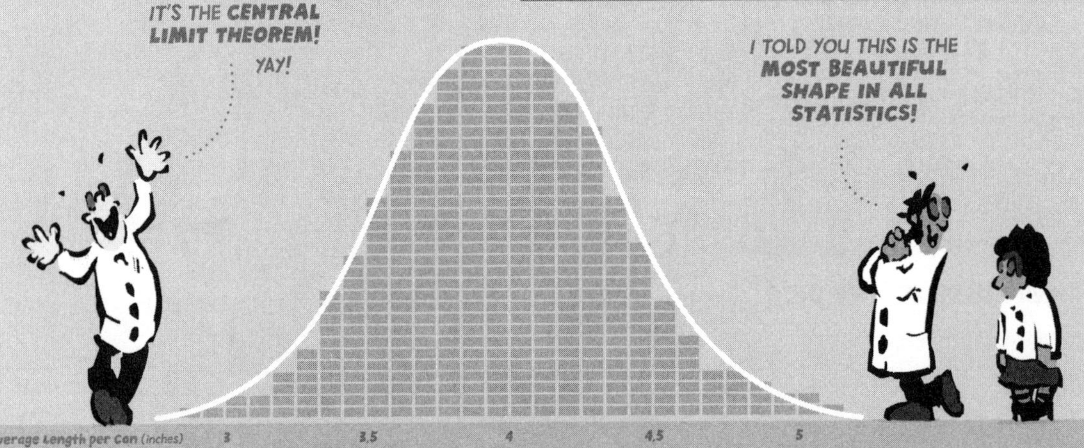

MORE SPECIFICALLY, BECAUSE THE PILE IS **NORMAL-SHAPED**...

... WE **ONLY** NEED TO KNOW ITS **CENTER VALUE** AND **STANDARD DEVIATION** TO MAKE THAT FANCY MATH WORK!*

* IF YOU LIKE MATH, SEE **PAGE 219** FOR MORE DETAILS.

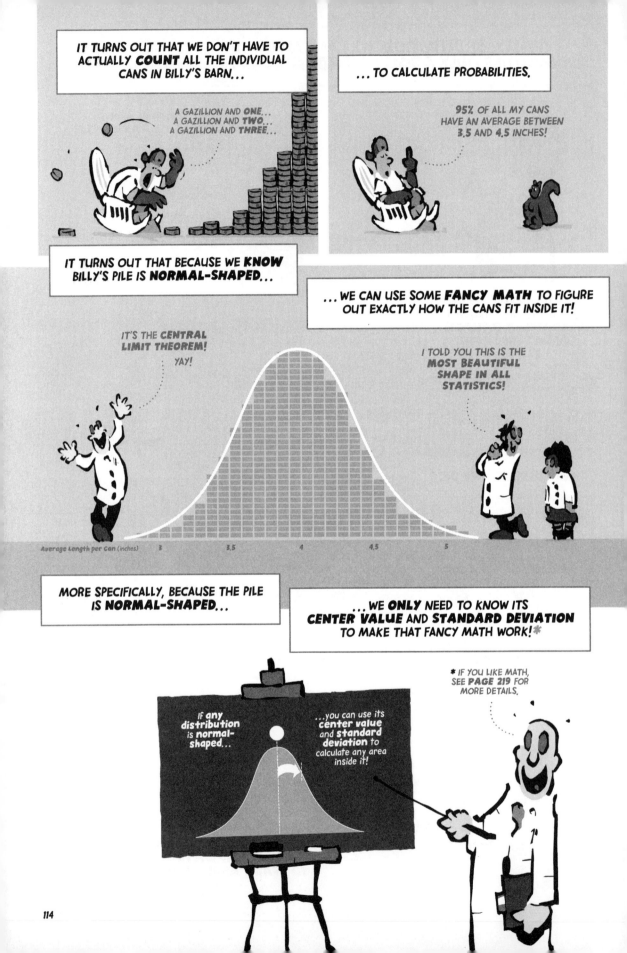

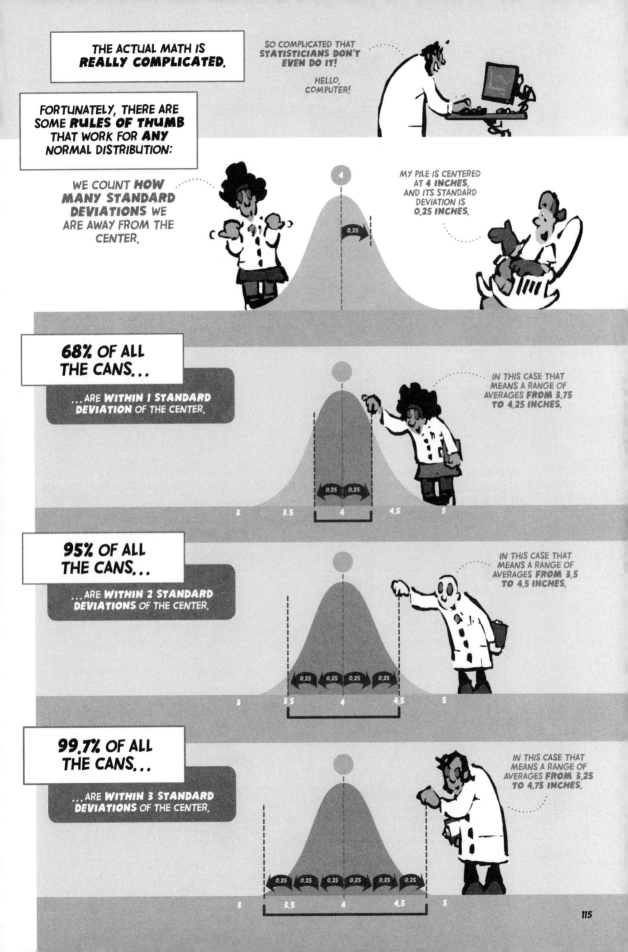

THE ACTUAL MATH IS **REALLY COMPLICATED.**

SO COMPLICATED THAT **STATISTICIANS DON'T EVEN DO IT!**
HELLO, COMPUTER!

FORTUNATELY, THERE ARE SOME **RULES OF THUMB** THAT WORK FOR **ANY** NORMAL DISTRIBUTION:

WE COUNT **HOW MANY STANDARD DEVIATIONS** WE ARE AWAY FROM THE CENTER.
4
0,25

MY PILE IS CENTERED AT **4 INCHES,** AND ITS STANDARD DEVIATION IS **0,25 INCHES.**

68% OF ALL THE CANS...

...ARE **WITHIN 1 STANDARD DEVIATION** OF THE CENTER.

95% OF ALL THE CANS...

...ARE **WITHIN 2 STANDARD DEVIATIONS** OF THE CENTER.

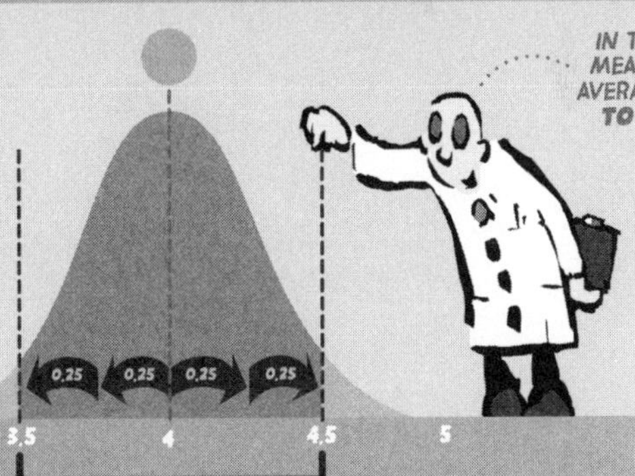

99.7% OF ALL THE CANS...

...ARE **WITHIN 3 STANDARD DEVIATIONS** OF THE CENTER.

IF ALL THOSE NUMBERS FEEL **BEWILDERING...**

...JUST FOCUS ON THE **SHADED AREAS.**

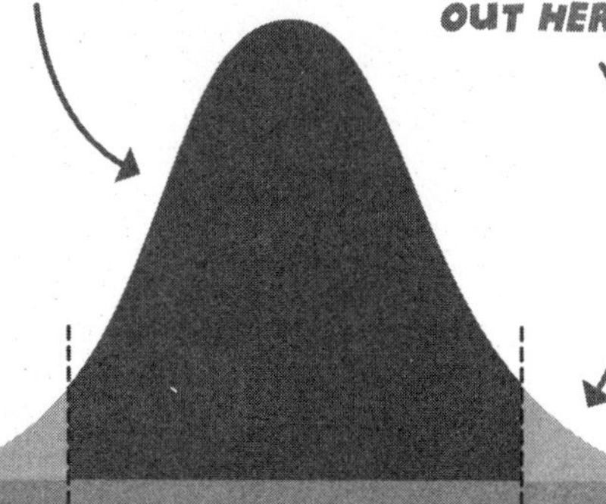

THE THING TO REMEMBER IS THAT THE SHADED AREAS **INSIDE BILLY'S SAMPLING DISTRIBUTION...**

...TRANSLATE DIRECTLY TO OUR CHANCES OF GRABBING AVERAGES **FROM THE SWAMP!**

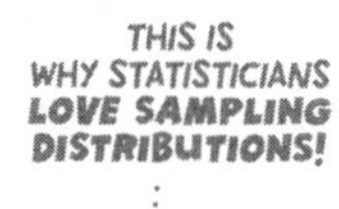

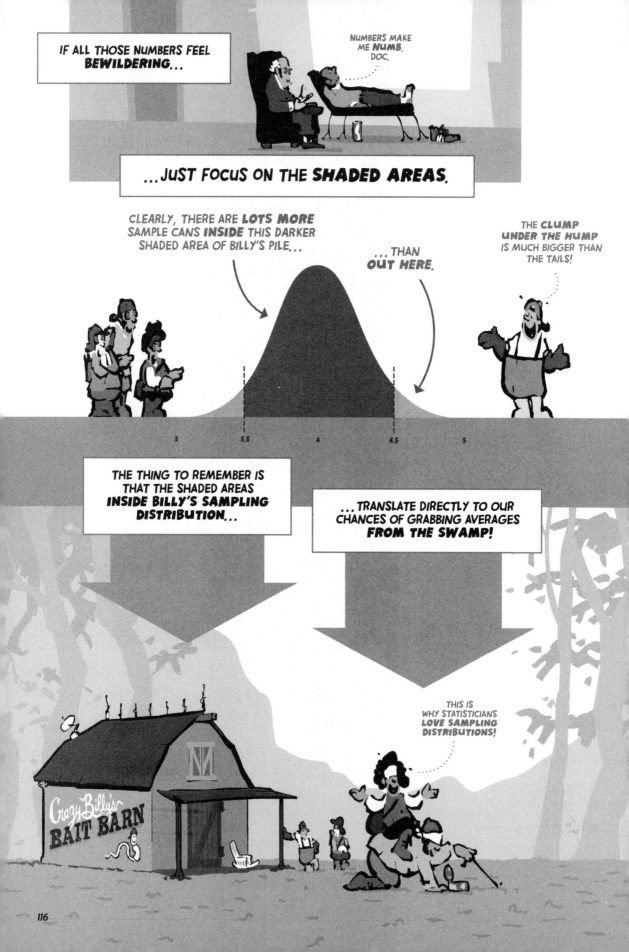

LET'S **RECAP:**

THE **FIRST COOL THING** ABOUT BILLY'S SAMPLING DISTRIBUTION...

...IS THAT IT SHOWS US **THE POPULATION AVERAGE!**

WHAT'S THE **AVERAGE LENGTH OF ALL THE WORMS IN YOUR SWAMP**, BILLY?

THE ANSWER'S **IN MY BAIT BARN!**

THE **SECOND COOL THING** ABOUT BILLY'S SAMPLING DISTRIBUTION...

...IS THAT WE CAN USE IT TO **CALCULATE PROBABILITIES ABOUT THE OVERALL POPULATION.**

AND BECAUSE WE KNOW IT'S NORMAL...

...ALL WE NEED TO KNOW ARE ITS **CENTER VALUE**...

...AND **STANDARD DEVIATION!**

IF WE GO GRAB ANOTHER RANDOM SAMPLE OF **30 WORMS** FROM THE SWAMP...
...HOW LIKELY IS IT TO HAVE AN AVERAGE BETWEEN **3.75** AND **4.25** INCHES?

LET ME **PEER INTO MY BAIT BARN** AND TELL YOU!

CLEARLY, IF WE WERE OUT **HUNTING THE POPULATION AVERAGE**...

...A SAMPLING DISTRIBUTION LIKE THE ONE IN **CRAZY BILLY'S BAIT BARN**...

...WOULD BE **INCREDIBLY USEFUL** TO US.

UNFORTUNATELY...
WHA?!
IT'S EMPTY!
...IT DOESN'T EXIST.

IN REALITY, WE NEVER HAVE AN ACTUAL SAMPLING DISTRIBUTION TO LOOK AT.
IT'S ALL IN MY HEAD!
I REMEMBER EVERY CAN I'VE EVER SOLD.
IN REALITY...
...ALL WE EVER GET IS ONE CAN.

&#@%!

SO WHAT'S THE **AVERAGE LENGTH OF ALL THE WORMS IN THE SWAMP?!**
UM...

CHAPTER 9
INFERENCE

OBVIOUSLY, WE **STILL HAVE A PROBLEM...**

...AND IT BOILS DOWN TO **THIS:**

WE'RE HUNTING FOR SOMETHING THAT WE **CAN'T LOOK AT DIRECTLY.**

THERE'S **NO WAY** TO LOOK INTO ONE SAMPLE...

...AND **SEE THE POPULATION AVERAGE.**

WE'RE JUST **30 WORMS,** AND THERE ARE A GAZILLION MORE IN THE SWAMP.

IT'S AS IF WE'RE GROPING THROUGH THE MIST, **HUNTING FOR BIGFOOT.**

I **BELIEVE** WITH ALL MY HEART THAT HE IS OUT THERE!

AND YET YOU'LL **NEVER** BE ABLE TO ACTUALLY FIND HIM.

FORTUNATELY, ALTHOUGH WE **CAN'T ACTUALLY SEE** THE THING WE'RE HUNTING FOR...

ALL I CAN SEE IS **FOG.**

...WE **CAN** STILL SEARCH FOR **CLUES...**

...TO HELP US GUESS WHERE IT **MIGHT** BE.

WHEN WE MAKE A **GUESS** ABOUT THE **WHEREABOUTS OF THE POPULATION AVERAGE...**

IT'S GOT TO BE AROUND HERE SOMEWHERE...

...WE CAN **BASE OUR GUESS** ON **SOMETHING WE'RE CONFIDENT ABOUT...**

...I'D PUT MONEY ON IT.

...AND WE'VE **ALREADY LEARNED** WHAT **THAT IS:**

IN THE LONG RUN, RANDOM SAMPLE AVERAGES TEND TO CLUSTER AROUND THE POPULATION AVERAGE...
...IN THIS BEAUTIFUL SHAPE!

IT'S THE CENTRAL LIMIT THEOREM!
YAY!

HERE'S HOW TO **THINK ABOUT** WHAT WE'RE ABOUT TO DO:

STATISTICIANS CALL THIS PROCESS **INFERENCE...**

...AND WE'RE GOING TO SPEND THE REST OF THIS CHAPTER OUTLINING **THE FIRST STEP.**

OUR **BASIC TASK** AT THIS STAGE...

...IS TO **DRAW A PICTURE.**

IT'S A PICTURE OF HOW WE THINK **SAMPLE AVERAGES WOULD PILE UP...**

...**IF** WE WENT OUT AND **GATHERED GAZILLIONS OF THEM.**

REMEMBER, WE DRAW THIS THING...

...BECAUSE WE'RE **HUNTING FOR THE POPULATION AVERAGE.**

AND IT'S ALL BASED ON INFORMATION WE GET FROM OUR **ONE RANDOM SAMPLE.**

IT'S LIKE WE'RE MAKING OUR **BEST POSSIBLE GUESS** AT WHAT CRAZY BILLY'S IMAGINARY PILE LOOKS LIKE...

...USING **ONLY ONE CAN.**

THIS MIGHT **SEEM LIKE MAGIC...**

...BUT IT'S ACTUALLY **PRETTY SIMPLE.**

1. REMEMBER, SOME **RESTRICTIONS APPLY.** SEE PAGE 102.

HOWEVER, BECAUSE WE **DON'T ACTUALLY KNOW THE REAL POPULATION VALUES**...

...WE SIMPLY **REPLACE THEM** WITH VALUES WE GRAB **FROM OUR SAMPLE.**

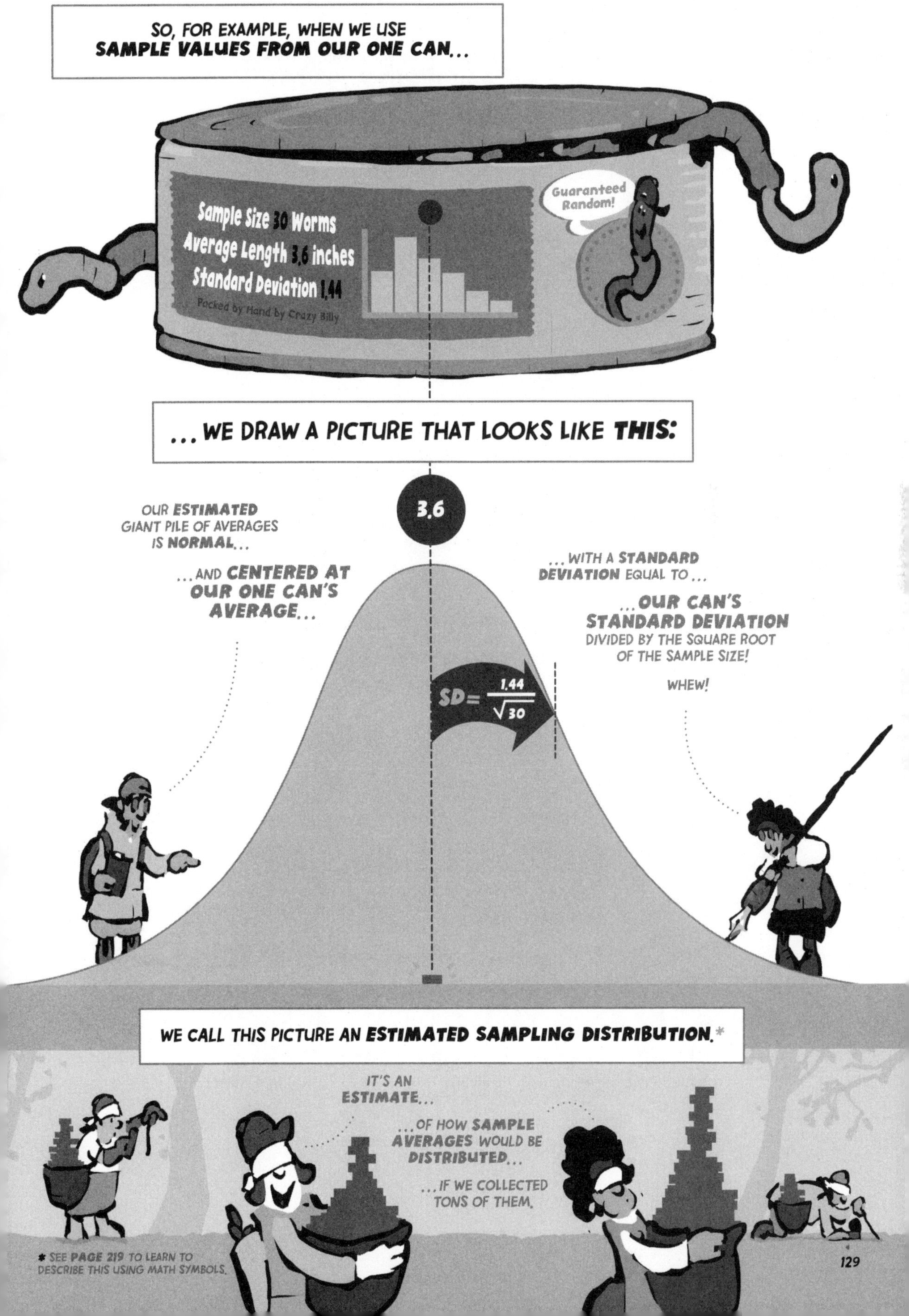

SO, FOR EXAMPLE, WHEN WE USE **SAMPLE VALUES FROM OUR ONE CAN...**

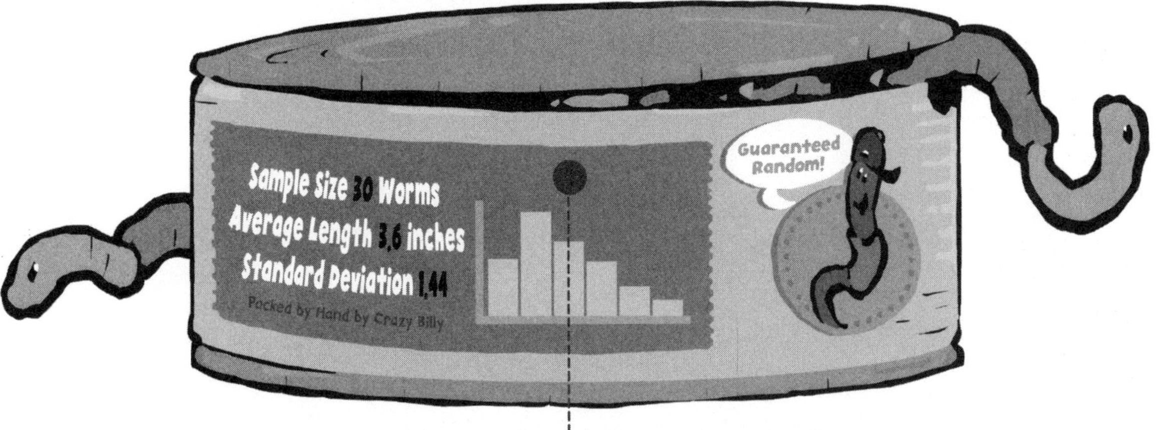

...WE DRAW A PICTURE THAT LOOKS LIKE **THIS:**

3.6

OUR **ESTIMATED** GIANT PILE OF AVERAGES IS **NORMAL...**

...AND **CENTERED AT OUR ONE CAN'S AVERAGE...**

...WITH A **STANDARD DEVIATION** EQUAL TO...

...**OUR CAN'S STANDARD DEVIATION** DIVIDED BY THE SQUARE ROOT OF THE SAMPLE SIZE!

WHEW!

$$SD = \frac{1.44}{\sqrt{30}}$$

WE CALL THIS PICTURE AN **ESTIMATED SAMPLING DISTRIBUTION.***

IT'S AN ESTIMATE...
...OF HOW SAMPLE AVERAGES WOULD BE DISTRIBUTED...
...IF WE COLLECTED TONS OF THEM.

* SEE **PAGE 219** TO LEARN TO DESCRIBE THIS USING MATH SYMBOLS.

NOW THAT WE'VE USED OUR **ONE SAMPLE...**

...TO BUILD AN **ESTIMATED SAMPLING DISTRIBUTION...**

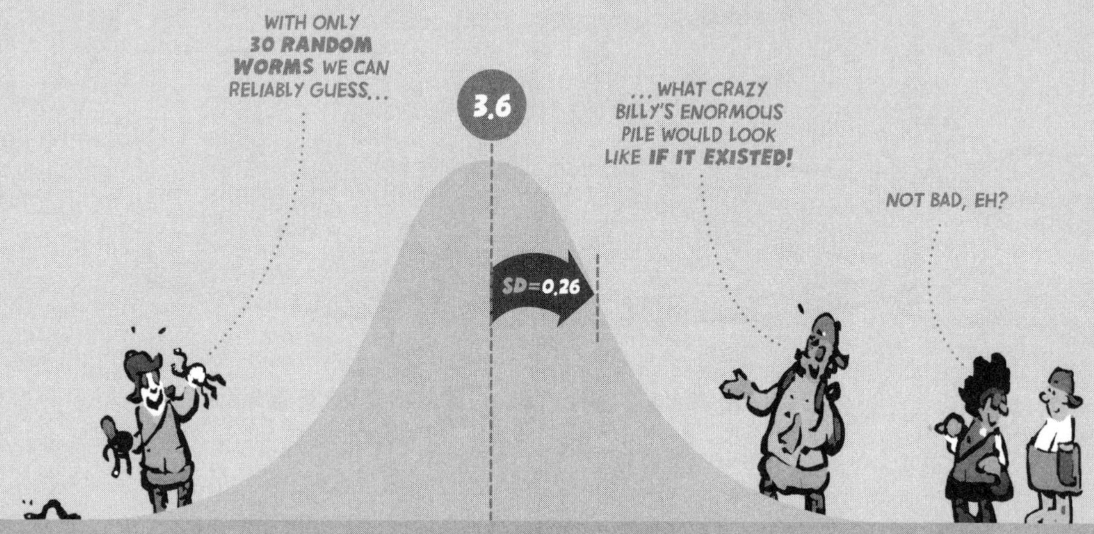

...WE CAN RESUME OUR **HUNT FOR THE POPULATION AVERAGE.**

CHAPTER 10
CONFIDENCE

REMEMBER THAT OUR ULTIMATE GOAL IS TO **LEARN SOMETHING ABOUT THE POPULATION AVERAGE.**

UNFORTUNATELY, DESPITE THE **MAGIC TRICKS** WE JUST LEARNED...

...WE WILL STILL **NEVER** BE ABLE TO **PINPOINT IT DIRECTLY...**

...**EVER!**

THAT'S WHY WE'RE LEARNING TO **DO INFERENCE.**

SO FAR WE'VE COVERED THE **FIRST STEP** IN THE **INFERENCE PROCESS**...
CAREFULLY, LOVINGLY DRAW YOUR ESTIMATED SAMPLING DISTRIBUTION...
...WITH EXQUISITE ATTENTION TO THE IMPORTANT AND CRUCIAL DETAILS.

...BUT WE STILL NEED TO COVER **THE SECOND.**
NOW CHOP IT INTO BITS!

SO IN THIS CHAPTER WE'RE GOING TO LEARN TO **REFINE OUR DRAWING**...
...GIVE IT A LITTLE **TRIM AROUND THE EDGES**...
...AND USE WHAT'S LEFT TO **CALCULATE OUR CONFIDENCE.**

MUCH BETTER!
NOW I CAN BE **CONFIDENT** ABOUT WHAT I'M LOOKING AT.

SINCE WE ALREADY KNOW **HOW TO DRAW AN ESTIMATED SAMPLING DISTRIBUTION...**

ISN'T SHE **BEAUTIFUL?**

...LEARNING HOW TO **CALCULATE** OUR **CONFIDENCE** IS EASY.

WE SIMPLY **PEER DOWN INTO** THE THING WE JUST DREW...

THIS TIME LET'S MEASURE **2 STANDARD DEVIATIONS AWAY FROM THE CENTER VALUE...**
...ON EITHER SIDE.

...AND **CHOP OFF ITS TAILS!**

SNIP SNIP!
HOLD STILL, THIS WON'T HURT A BIT.

THEN **WE MAKE A STATEMENT LIKE THIS ONE:**

WE'RE **95% CONFIDENT...**

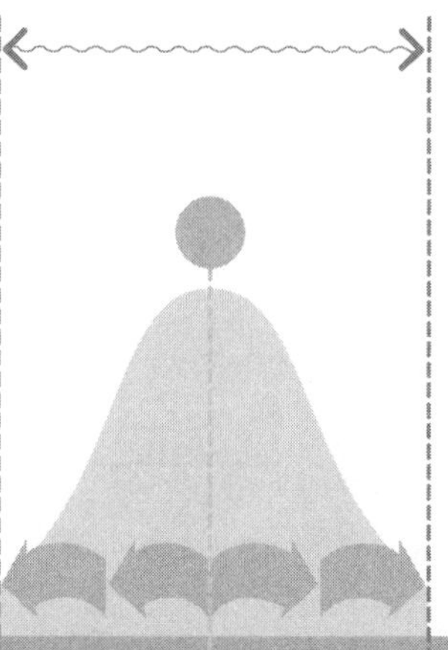

...THAT THE POPULATION AVERAGE IS SOMEWHERE INSIDE THIS RANGE!

YAY!
WE'LL RETURN TO THESE BITS IN **CHAPTER 11.**

THAT'S **ALL IT TAKES.**

TECHNICALLY, IF WE WANT TO BE **MORE CONFIDENT**, WE CAN **CHOP FARTHER OUT.**

IF WE WANT TO BE **99.7% CONFIDENT**...

...WE MEASURE AND CHOP **3 SDs** AWAY FROM THE CENTER...

...ON EITHER SIDE.

AND IF WE WANT TO BE **LESS CONFIDENT**, WE CAN **CHOP CLOSER IN.**

IF WE WANT TO BE **68% CONFIDENT**...

...WE MEASURE AND CHOP **1 SD** AWAY FROM THE CENTER...

...ON EITHER SIDE.

IT'S BASED ON THE RULES WE LEARNED ON **PAGE 115!**

BUT WHEREVER WE CHOP, WE **ALWAYS** DECLARE OUR CONFIDENCE WITH A **TWO-PART STATEMENT**, LIKE THIS...

WE'RE 95% CONFIDENT... | ...THAT THE POPULATION AVERAGE IS SOMEWHERE INSIDE **THIS** RANGE! |

...THAT COMBINES A PARTICULAR **LEVEL OF CONFIDENCE...**

...WITH A **CONFIDENCE INTERVAL.***

* TECHNICALLY, A CONFIDENCE INTERVAL IS A KIND OF **INTERVAL ESTIMATE.** SEE **PAGE 220** FOR MORE.

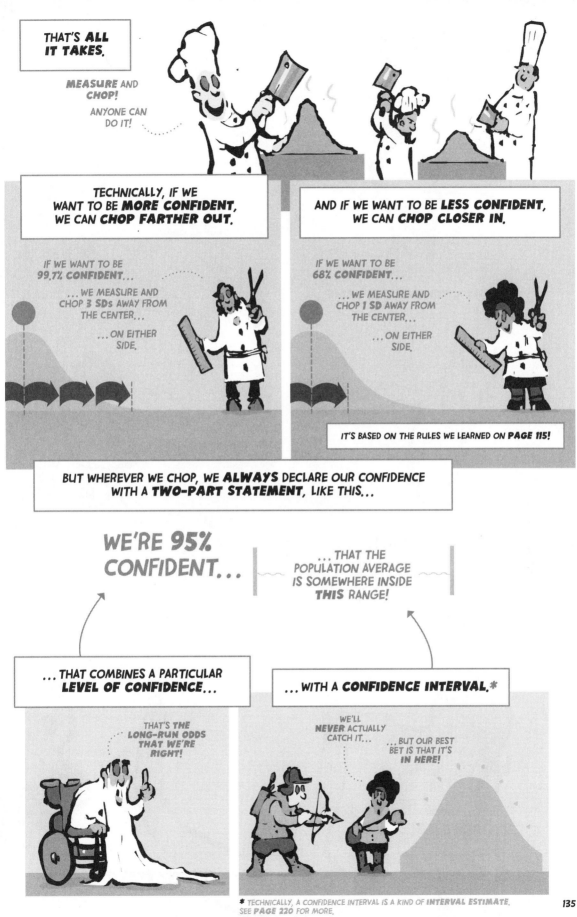

FOR EXAMPLE, IF WE TAKE THE ESTIMATED SAMPLING DISTRIBUTION WE BUILT WITH OUR **ONE CAN OF WORMS...**

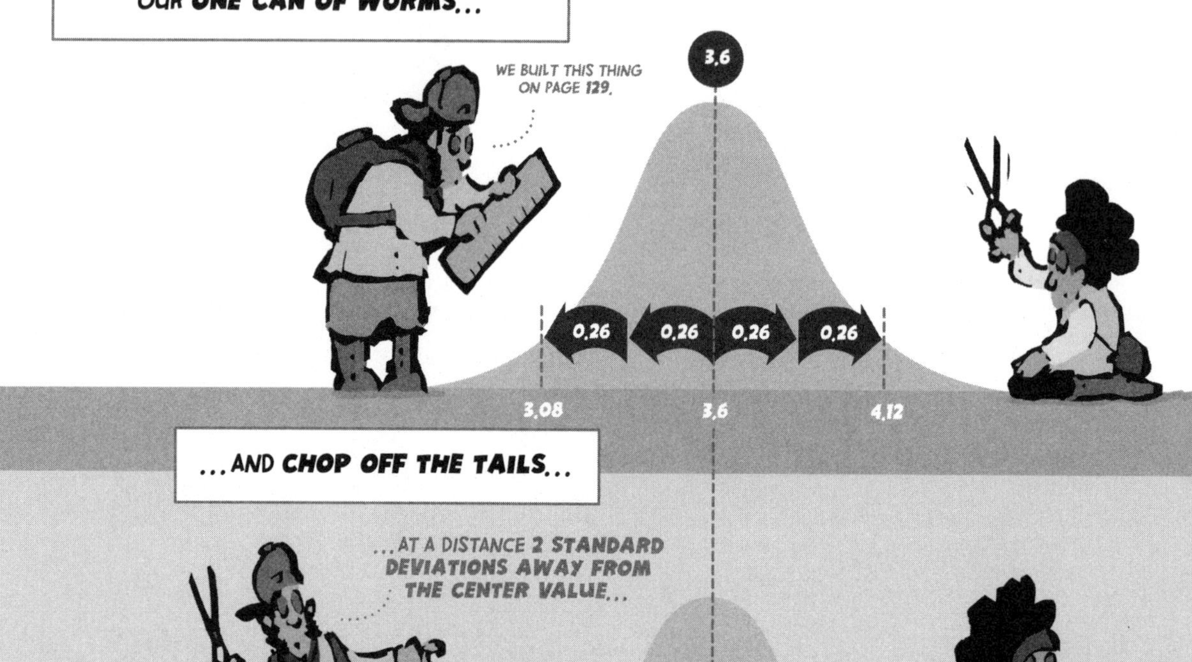

WE BUILT THIS THING ON PAGE 129.
3.6
0.26
0.26
0.26
0.26
3.08
3.6
4.12

...AND **CHOP OFF THE TAILS...**

...AT A DISTANCE 2 STANDARD DEVIATIONS AWAY FROM THE CENTER VALUE...
3.08
4.12

...WE CAN SAY **THIS:**

WE'RE 95% CONFIDENT...
...THAT THE POPULATION AVERAGE IS BETWEEN 3.08 AND 4.12 INCHES!

BUT WHAT **EXACTLY DOES IT MEAN?**

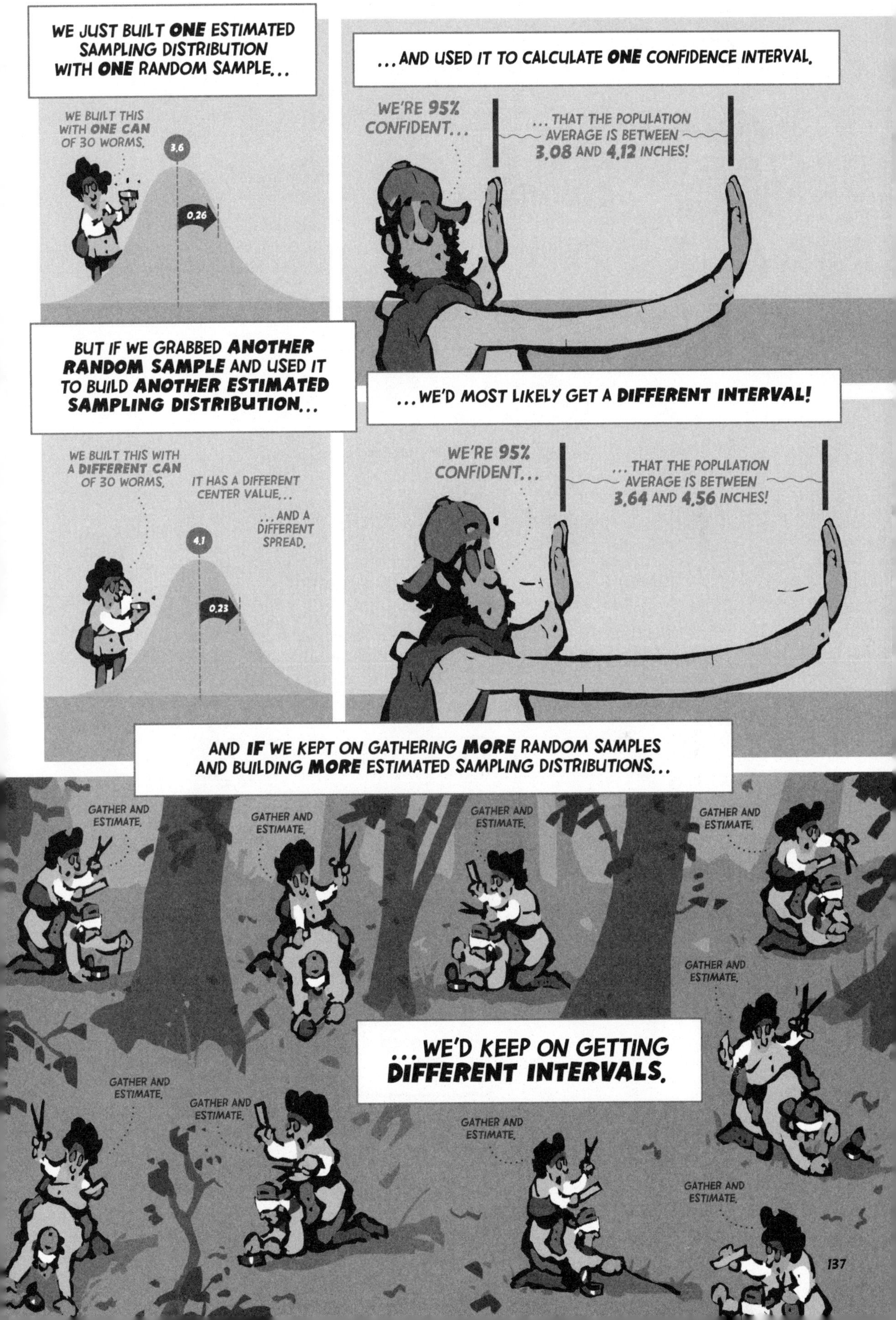

WE JUST BUILT **ONE** ESTIMATED SAMPLING DISTRIBUTION WITH **ONE** RANDOM SAMPLE...

...AND USED IT TO CALCULATE **ONE** CONFIDENCE INTERVAL.

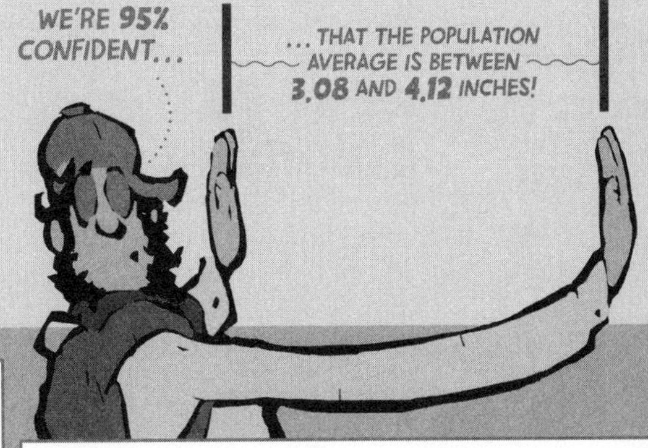

BUT IF WE GRABBED **ANOTHER RANDOM SAMPLE** AND USED IT TO BUILD **ANOTHER ESTIMATED SAMPLING DISTRIBUTION**...

...WE'D MOST LIKELY GET A **DIFFERENT INTERVAL!**

AND **IF** WE KEPT ON GATHERING **MORE** RANDOM SAMPLES AND BUILDING **MORE** ESTIMATED SAMPLING DISTRIBUTIONS...

GATHER AND ESTIMATE.

GATHER AND ESTIMATE.

GATHER AND ESTIMATE.

GATHER AND ESTIMATE.

GATHER AND ESTIMATE.

...WE'D KEEP ON GETTING **DIFFERENT INTERVALS.**

GATHER AND ESTIMATE.

GATHER AND ESTIMATE.

GATHER AND ESTIMATE.

GATHER AND ESTIMATE.

THIS IS IMPORTANT BECAUSE THE **ONLY THING** THAT THIS ACTUALLY MEANS...

...IS THAT **IF** WE CALCULATED A GAZILLION DIFFERENT RANGES **IN THIS SAME WAY**...

...ABOUT 19 OUT OF EVERY 20 **WOULD** INCLUDE THE TRUE POPULATION AVERAGE...

19 OUT OF 20 IS 95%!

...AND ABOUT 1 OUT OF EVERY 20 **WOULDN'T.**

1 OUT OF 20 IS **COMPLETELY WRONG!**

LET'S HOPE WE DIDN'T GRAB **THAT** CAN.

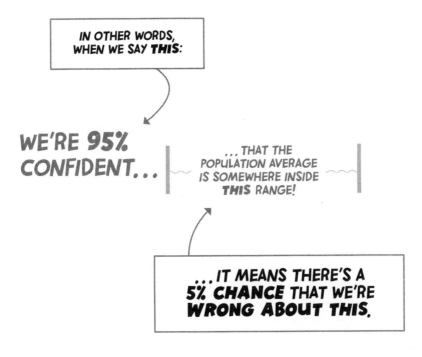

IN OTHER WORDS, WHEN WE SAY **THIS**:

WE'RE 95% CONFIDENT... ...THAT THE POPULATION AVERAGE IS SOMEWHERE INSIDE **THIS** RANGE!

...IT MEANS THERE'S A **5% CHANCE** THAT WE'RE **WRONG ABOUT THIS.**

IN WHICH CASE, THE POPULATION AVERAGE IS REALLY SOMEWHERE ELSE...
...AND WE MISSED THE MARK ENTIRELY!

THE **SAD TRUTH** IS THAT **ANY ONE SAMPLE** WE ASSEMBLE RANDOMLY FROM A POPULATION...

...**MIGHT** BE MISLEADING.

AND **IF OUR ONE SAMPLE** IS MISLEADING...

...THE ESTIMATED SAMPLING DISTRIBUTION WE BUILD ON TOP OF IT **WILL BE MISLEADING, TOO.**

IT'S A **SERIOUS CONCERN**...

...BUT WE CAN GET AROUND IT BY **ALWAYS** KEEPING THE **BIGGER PICTURE** IN MIND.

IN THE LONG RUN, DESPITE THE FACT THAT OUR ONE SAMPLE MIGHT BE MISLEADING...

...IT PROBABLY ISN'T...

...BECAUSE MOST RANDOM SAMPLE AVERAGES TEND TO CLUSTER AROUND THE POPULATION AVERAGE!

IN OTHER WORDS, OUR ONE CAN MIGHT RANDOMLY HAVE AN AVERAGE VALUE WAY OUT HERE...

WEIRDLY SHORT WORMS.

...OR WAY OUT HERE...

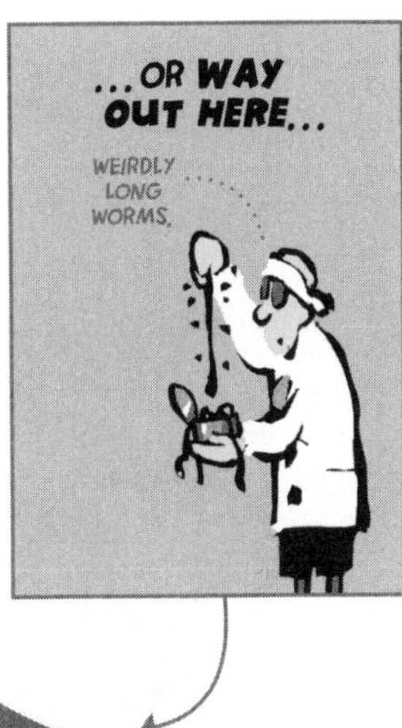

...BUT IT'S UNLIKELY...

...AND WE CAN BE CONFIDENT ABOUT THAT.

IN SUM, UNDERSTANDING **STATISTICAL CONFIDENCE**...

...INVOLVES KEEPING OUR EYES ON THE LONG RUN **AND** THE SHORT RUN **AT THE SAME TIME.**

IN THE LONG RUN, OUR ESTIMATION AND CHOPPING CALCULATIONS WORK, **PERIOD**...

...BUT **IN THE SHORT RUN**, IT'S ALWAYS POSSIBLE THAT WE **GRABBED A DUD!**

CHAPTER 11
THEY HATE US

GRRRRR.
THEY WANT TO KILL US!
HOW CONFIDENT ARE YOU ABOUT THAT?

WHEN WE USE ONE **SAMPLE**...

...TO **CALCULATE OUR STATISTICAL CONFIDENCE** ABOUT AN ENTIRE **POPULATION**...

...WE'RE DOING SOMETHING **AMAZING!**

WE'RE MAKING A **GUESS**...

...A **DEPENDABLE** GUESS...

...ABOUT **SOMETHING WE CAN'T SEE, BUT CAN ONLY IMAGINE.**

YOO HOO...
...ARE THERE MORE OF YOU DOWN THERE?
HE'LL NEVER KNOW FOR CERTAIN.

IT ALL STARTS WITH ONLY THREE NUMBERS.
A REASONABLY LARGE SAMPLE SIZE...
...A SAMPLE AVERAGE...
...AND A SAMPLE STANDARD DEVIATION.
BUT REMEMBER, IT ONLY WORKS IF YOU GATHER YOUR SAMPLE MERMAIDS RANDOMLY.

WITH ONLY THESE THREE NUMBERS WE CAN BUILD AN ESTIMATED SAMPLING DISTRIBUTION...
WHICH IS LIKE IMAGINING HOW A GAZILLION SAMPLES WOULD LOOK...
...IF WE PILED THEM UP BY THEIR AVERAGE VALUE.
8.5
1.25
WE KNOW THE PILE WOULD BE NORMAL-SHAPED IF THE SAMPLE SIZE IS LARGE ENOUGH.
PLUS, WE CAN ESTIMATE ITS CENTER VALUE AND STANDARD DEVIATION WITH OUR SAMPLE VALUES.
5.5 6 6.5 7 7.5 8 8.5 9 9.5 10 10.5 11 11.5
...AND CHOP OFF ITS TAILS...
WE COUNT OUTWARD BY ITS STANDARD DEVIATIONS...
...TO LEAVE THIS MIDDLE CLUMP WHOSE PROBABILITY VALUE WE KNOW PRECISELY.
7
10
...TO GET A SINGLE DEPENDABLE STATEMENT...
IN THE LONG RUN, THIS WORKS!
...THAT COMBINES A LEVEL OF CONFIDENCE...
...WITH A CONFIDENCE INTERVAL.
I'M 95% CONFIDENT...
...THAT THE POPULATION AVERAGE OF ALL THE MERMAIDS IN THE ENTIRE LAGOON IS SOMEWHERE BETWEEN 7 AND 10 CENTIMETERS!

AS WE'VE LEARNED, BECAUSE THIS PROCESS INVOLVES A BUNCH OF **MATH**...

AAHHHHH!

...IT WORKS ONLY ON **QUALITIES** THAT WE **CAN MEASURE WITH NUMBERS.**

SO IT **MIGHT SEEM** LIKE IT **WON'T WORK** ON QUALITIES THAT **AREN'T OBVIOUSLY NUMERICAL**...

ARE THEY **HAPPY**?

DO THEY ALL **SING BEAUTIFULLY?**

IF I POKE THEM WITH A STICK, **HOW MUCH DO THEY HURT?**

...BUT THAT'S **NOT ALWAYS THE CASE.**

THE TRUTH IS, WE CAN CALCULATE OUR CONFIDENCE ABOUT **ANY QUALITY**...

ARE MERMAIDS **OPTIMISTIC?**
HOW **SMART** ARE THEY?
DO THEY ENJOY EATING **SUSHI?**

...**IF** WE CAN FIGURE OUT A WAY TO **MEASURE THAT QUALITY**...

...AND ARRANGE IT ON A **NUMBER LINE.**

IF YOU SCORED OVER HERE, YOU'RE AN **IDIOT!**
IF YOU SCORED OVER HERE, YOU'RE A **GENIUS!**
Score:
60
80
100
120
140

IN THIS CHAPTER WE'RE GOING TO DO JUST THIS SORT OF THING...

...TO INVESTIGATE A QUESTION ABOUT **HATRED.**

HOW DO I **HATE THEE?**
LET ME **COUNT THE WAYS**...

EVERYONE KNOWS THAT THE AWFUL WRETCHES ON THE PLANET OF **BLIP**...

...**HATE** THE GOOD PEOPLE ON THE NEIGHBORING PLANET OF **BLEEEP**.

NYA NYA!

YOU @*$&-ING BLEEEPS!

THEY **HATE** US!

THEY'VE HATED US FOR A **TRILLION GENERATIONS**.

AND OUR **QUESTION** IS...

...**IS THIS TRUE?**

BECAUSE WE CAN'T POSSIBLY ASK ALL THE **785,000,000,000 BLIPS** WHO LIVE ON THAT PLANET HOW THEY FEEL...

...WE HAVE **NO CHOICE** BUT **TO BASE OUR ANSWER** ON A **RANDOM SAMPLE** OF THEM.

THAT'S A LOT OF BLIPS.

IN OTHER WORDS, **WE HAVE TO USE STATISTICS**.

REMEMBER, **NOBODY** USES STATISTICS **UNLESS THEY HAVE TO!**

HOWEVER, BEFORE WE CAN GO OUT AND **GATHER A RANDOM SAMPLE...**

...WE HAVE TO FIGURE OUT A WAY TO TRANSLATE **HOW EACH BLIP FEELS ABOUT BLEEEPS...**

...INTO A **NUMBER.**

IN THIS CASE WE CAN **INVENT A SCALE...**

...TO PURE **LOVE.**

SO, FOR EACH BLIP WE INTERVIEW...

...WE'LL GET **A SINGLE NUMBER BETWEEN -10 AND 10.**

THE NEXT STEP IS TO PAINSTAKINGLY **GATHER OUR SAMPLE DATA**...

ON A SCALE OF -10 TO 10, HOW DO **YOU** FEEL ABOUT BLEEEPS?

ON A SCALE OF -10 TO 10, HOW DO **YOU** FEEL ABOUT BLEEEPS?

ON A SCALE OF -10 TO 10, HOW DO **YOU** FEEL ABOUT BLEEEPS?

...**RANDOMLY**...

THERE CAN BE **NO SYSTEMATIC DIFFERENCE**...

...BETWEEN THE SAMPLE OF BLIPS WE INTERVIEW...

...AND ANY OTHER SAMPLE WE MIGHT GRAB...

...SO LET'S GRAB OUR SAMPLE BLIPS RANDOMLY **FROM ALL OVER THEIR PLANET**...

...FROM **HERE**...

...**HERE**...

...BACK **HERE**...

...**HERE**...

...**HERE**...

...**HERE**...

...ETC...

...ETC...

...KEEPING IN MIND THAT WE HAVE TO **GATHER ENOUGH OF IT**...

...TO **MAKE SURE THE MATH WORKS.**

...WE WON'T STOP UNTIL WE'VE INTERVIEWED **100 OF THEM.**

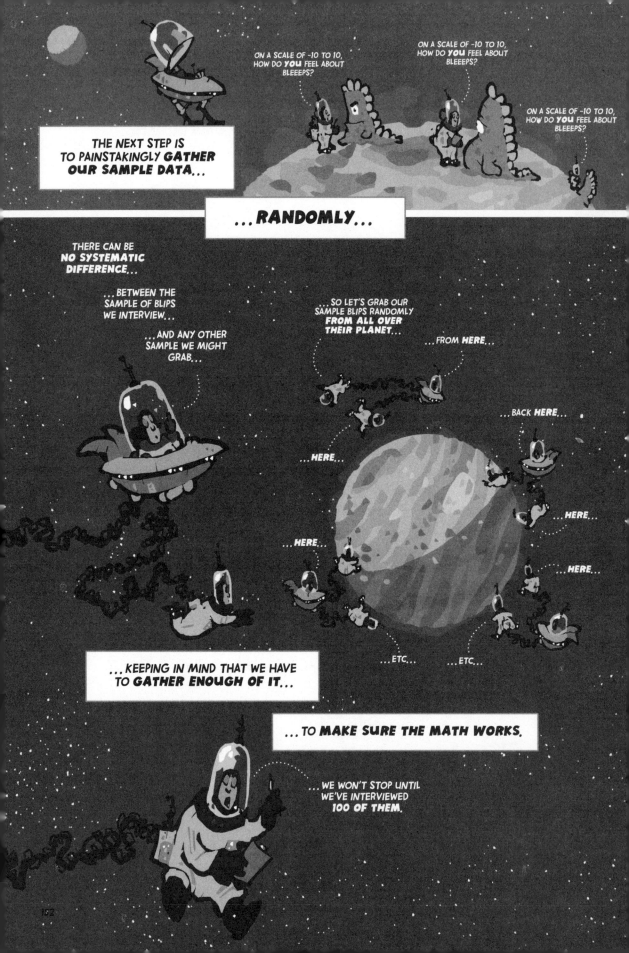

THEN, ONCE WE'VE ASSEMBLED OUR **SAMPLE DATA...**

OVERALL, OUR DATA IS SLIGHTLY **SHIFTED LEFT OF ZERO....**

...AND **SKEWED TO THE RIGHT.**

THESE FIVE BLIPS TRULY **HATE** BLEEEPS.

ONE BLIP REPORTED **LOVING** BLEEEPS!

Blip Frequency

20

15

10

Sentiment Toward Bleeeps

-10 -5 0 5 10

...WE CAN **PONDER IT A BIT...**

CLEARLY, LOTS OF OUR SAMPLE BLIPS FEEL **NEGATIVELY** ABOUT BLEEEPS...

...BUT DOES THAT MEAN THE **WHOLE PLANET** IS **SEETHING WITH HATRED?**

...AS WE EXTRACT THE **THREE NUMBERS** THAT WE NEED TO DO **STATISTICAL INFERENCE.**

ONCE AGAIN, WE START WITH ONLY **THREE NUMBERS.**
IF WE CAN'T GET A **REASONABLY LARGE** SAMPLE SIZE...
...WE HAVE TO **TWEAK OUR TOOLS A BIT**, WHICH WE'LL LEARN HOW TO DO IN **CHAPTER 14.**
OUR REASONABLY LARGE **SAMPLE SIZE** IS 100...
...OUR **SAMPLE AVERAGE** IS -1...
...AND OUR **SAMPLE STANDARD DEVIATION** IS 4.

WITH THOSE THREE NUMBERS WE BUILD AN ESTIMATED SAMPLING DISTRIBUTION...
IT'S NORMAL...
...AND CENTERED AT OUR SAMPLE AVERAGE.
HURRAH!!
-1
WITH A STANDARD DEVIATION EQUAL TO...
...OUR SAMPLE'S STANDARD DEVIATION...
...DIVIDED BY THE SQUARE ROOT OF THE SAMPLE SIZE.
4 / √100
-2.0
-1.0
0.0
...AND CHOP OFF ITS TAILS...
WE'LL COUNT OUTWARD BY 2 STANDARD DEVIATIONS...
...TO GIVE US 95% OF THE AREA INSIDE THE HUMP.
THEN WE LOOK AT THE RANGE OF VALUES THAT COVERS.
0.4
0.4
0.4
0.4
...TO GET A SINGLE DEPENDABLE STATEMENT...
-1.8
-1
-0.2
THOUGH IN THE SHORT RUN, WE MIGHT BE WRONG.
...THAT COMBINES A LEVEL OF CONFIDENCE...
...WITH A CONFIDENCE INTERVAL.
WE'RE 95% CONFIDENT...
...THAT THE POPULATION AVERAGE OF ALL THE BLIPS ON THE ENTIRE PLANET IS SOMEWHERE BETWEEN -1.8 AND -0.2!

SO DO **BLIPS**...

...**HATE** BLEEEPS?

WE WANT TO **MANGLE YOUR EYEBALLS WITH OUR THUMBS**...

...**CUT OFF YOUR TAILS WITH A DULL KNIFE**...

...**REGURGITATE YOUR SPLEEN!**

...THEN LIGHT THE WHOLE THING **ON FIRE!**

SEE...?

THEY'RE A BUNCH OF **SPITEFUL EVILDOERS!**

SINCE WE CAN'T INTERVIEW THE ENTIRE POPULATION, WE'LL **NEVER KNOW FOR CERTAIN.**

BUT OUR RANDOM SAMPLE DATA...

...**STRONGLY** SUGGESTS THEY **DON'T**, ACTUALLY.

REMEMBER THAT THE **BEST WE CAN OFFER** WITH STATISTICS IS A **NUANCED PORTRAIT...**

...BECAUSE **ANY CONCLUSION BASED ON ONE SAMPLE...**

...MIGHT BE DEAD WRONG.

AND EVEN IF WE **EXPAND OUR LEVEL OF CONFIDENCE...**

...TO COVER A **WIDER** INTERVAL...

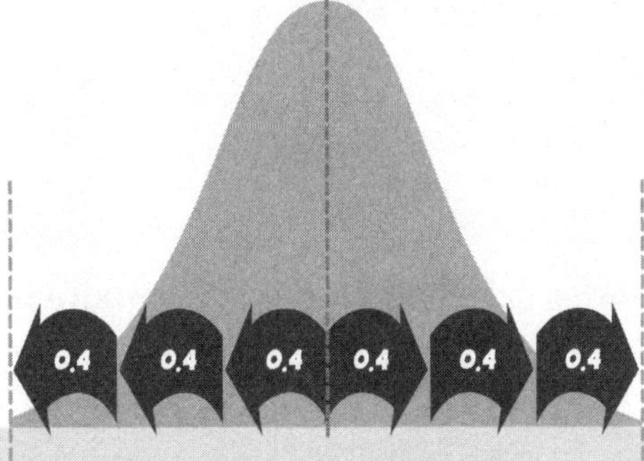

...WE STILL **MIGHT** BE WRONG.

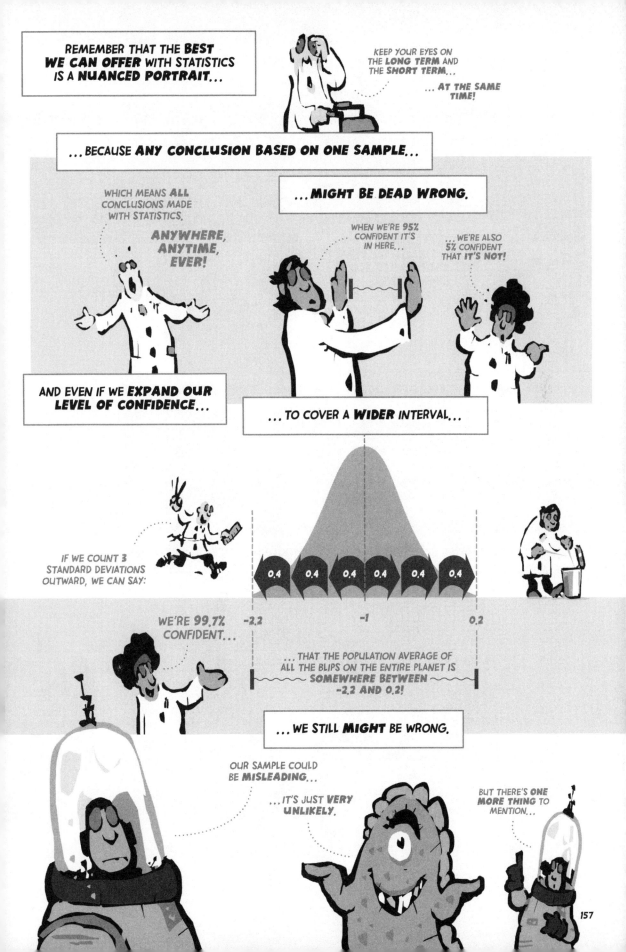

WE JUST CALCULATED **95%** AND **99.7%** CONFIDENCE INTERVALS...

...BASED ON ONE RANDOM SAMPLE OF **100 BLIPS**.

WE'RE **REALLY CONFIDENT**...
...THAT WE PROBABLY **DON'T LIKE YOU VERY MUCH!**

BUT THERE'S ONE OTHER THING WE COULD HAVE DONE TO **BUY OURSELVES MORE CONFIDENCE.**

YOU CAN **NEVER GET ENOUGH CONFIDENCE.**
MORE IS **ALWAYS BETTER!**

IF WE HAD STARTED THE WHOLE PROCESS BY INTERVIEWING **MORE RANDOM BLIPS**...

ON A SCALE OF -10 TO 10, HOW DO **YOU** FEEL ABOUT BLEEEPS?
LET'S INTERVIEW **MORE THAN 100!**
LET'S NOT STOP UNTIL WE'VE INTERVIEWED **225!**

...OUR ESTIMATED SAMPLING DISTRIBUTION WOULD HAVE BEEN **NARROWER**...

THE WHOLE PILE BECOMES **MORE SLENDER!**
WE PREDICTED THIS WOULD HAPPEN ON **PAGE 98.**

...AND THAT WOULD HAVE MADE OUR CONFIDENCE INTERVALS **NARROWER** AND THEREFORE **MORE PRECISE!**

TO SEE HOW IT WORKS, IMAGINE WE HAD ORIGINALLY STARTED WITH **THESE THREE NUMBERS.**

THE **SAMPLE SIZE IS 225.**

THAT'S MORE THAN TWICE AS MANY BLIPS.

THE **SAMPLE AVERAGE IS -1.**

THE **SAMPLE SD IS 4.**

IT'S **UNLIKELY** OUR SAMPLE SIZE AND SAMPLE SD WOULD **REALLY BE THE SAME WITH A LARGER SAMPLE...**

...BUT LET'S USE THE SAME NUMBERS SO WE CAN SEE THE EFFECTS OF CHANGING **ONLY THE SAMPLE SIZE.**

JUST LIKE BEFORE, WITH THESE THREE NUMBERS WE CAN **BUILD AN ESTIMATED SAMPLING DISTRIBUTION...**

IT'S NORMAL...

...AND CENTERED AT OUR SAMPLE AVERAGE...

...BUT THE WHOLE THING IS **NARROWER** THIS TIME BECAUSE OUR SAMPLE SIZE WAS LARGER.

-1
$\frac{4}{\sqrt{225}}$
0.26 0.26 0.26 0.26 0.26 0.26
-2.5 -2.0 -1.5 -1.0 -0.5 0.0 0.5

...AND **CHOP OFF ITS TAILS.**

LET'S COUNT 3 STANDARD DEVIATIONS OUTWARD TO GET A 99.7% CONFIDENCE INTERVAL.

AND THIS TIME **ANY PARTICULAR CONFIDENCE LEVEL** WE CHOOSE...

...WILL HAVE A **MUCH MORE PRECISE** INTERVAL.

THIS TIME WE'RE **99.7%** CONFIDENT...

...THAT THE POPULATION AVERAGE IS **SOMEWHERE BETWEEN -1.78 AND -0.22.**

THAT'S ABOUT THE SAME AS THE **95%** CONFIDENCE INTERVAL WE GOT WITH **100** BLIPS!

THIS, ULTIMATELY, IS THE REASON A **LARGER SAMPLE SIZE IS BETTER!**

IF YOU CAN **INCREASE YOUR SAMPLE SIZE...**

...YOU **SHOULD.**

IT'LL MAKE YOU **MORE CONFIDENT!**

IN THIS CHAPTER WE CREATED A **NUMERICAL SCALE** THAT DESCRIBES ***HOW ONE GROUP OF PEOPLE FEELS ABOUT ANOTHER.***

I'M A **10!**
. . . WILL YOU **MARRY ME?**

THIS SAME SORT OF TRICK CAN WORK TO INVESTIGATE **ALL KINDS OF DIFFERENT QUESTIONS...**

ON A SCALE OF **1** TO **10**. . .
. . . TELL ME HOW MUCH THIS **HURTS.**

. . . BECAUSE **IF** WE CAN GATHER ENOUGH **RANDOM NUMERICAL SAMPLE DATA...**

. . . WE CAN **CALCULATE OUR CONFIDENCE** ABOUT **ANY POPULATION...**

. . . THAT **LIES BEYOND OUR REACH.**

CHAPTER 12
HYPOTHESIS TESTING

IN THE LAST FEW CHAPTERS WE LEARNED HOW TO **CALCULATE OUR CONFIDENCE...**

...BY **BUILDING AN ESTIMATED SAMPLING DISTRIBUTION...**

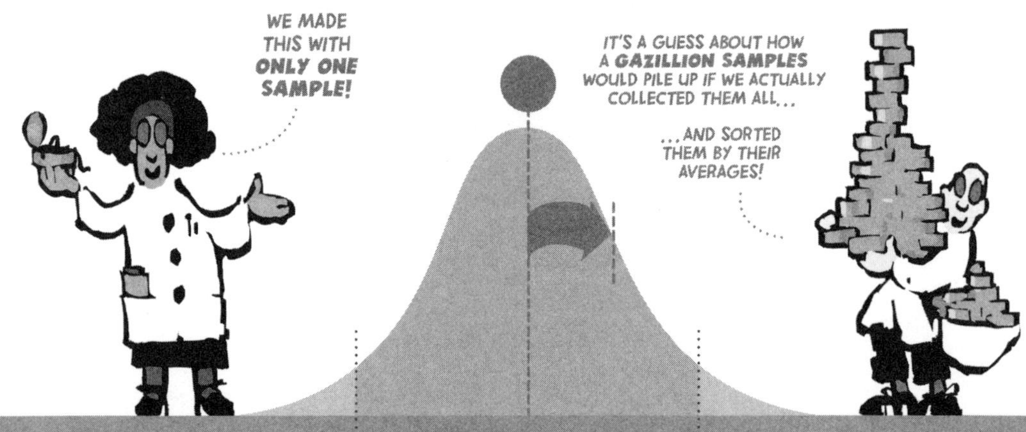

...AND CARVING OUT A **BIG CLUMP IN THE MIDDLE OF IT.**

IN THIS CHAPTER WE'RE GOING TO LEARN A **NEW TECHNIQUE.**

WE'RE GOING TO **TAKE OUR ESTIMATE...**

...AND SEE WHAT IT CAN TELL US WHEN WE **SLIDE IT TO A NEW CENTER LOCATION.**

IT'S ALL PART OF A PROCESS CALLED **HYPOTHESIS TESTING...***

THIS IS A **TEST...**

...OF WHETHER WE THINK THE POPULATION AVERAGE **MIGHT ACTUALLY REALLY BE...**

...RIGHT HERE!!

...ANOTHER **MAJOR STRATEGY** FOR **DOING INFERENCE.**

* SEE **PAGE 221** FOR A TECHNICAL DESCRIPTION OF THE STUFF WE COVER IN THIS CHAPTER.

HYPOTHESIS TESTING **SOUNDS FANCY**...

...BUT IT'S REALLY JUST ANOTHER WAY TO **HUNT FOR THE ELUSIVE POPULATION AVERAGE.**

AS WE'VE LEARNED, WE'LL **NEVER** BE ABLE TO SEE THE REAL POPULATION AVERAGE DIRECTLY...

...BUT IT TURNS OUT WE CAN MAKE SOME HEADWAY IN OUR QUEST BY **TAKING A STAB AT ITS PRECISE LOCATION.**

WHAT WE'LL ULTIMATELY BE DOING WITH **HYPOTHESIS TESTING**...

...IS **TESTING OUR GUESS**...

...BY **COMPARING IT** TO THE AVERAGE IN THE SAMPLE **WE ACTUALLY FOUND.**

BUT THE WHOLE PROCESS INVOLVES SEVERAL **TWISTS AND TURNS**...

...SO WE'RE GOING TO **TAKE IT SLOWLY.**

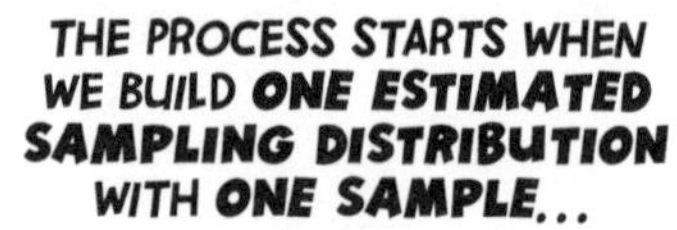
THE PROCESS STARTS WHEN WE BUILD **ONE ESTIMATED SAMPLING DISTRIBUTION** WITH **ONE SAMPLE**...

REMEMBER, THAT THING IS LIKE A PORTRAIT OF **HOW OTHER SAMPLE AVERAGES WOULD PILE UP**...
...**IF** THEY WERE CENTERED ABOVE THE AVERAGE IN OUR ONE SAMPLE.

...AND **SLIDE** IT OVER TO **ANOTHER LOCATION THAT WE'RE CURIOUS ABOUT.**
WE USUALLY HAVE A **PARTICULAR LOCATION** IN MIND...
...BUT WE'LL GET TO THAT IN THE NEXT CHAPTER.

THAT GIVES US A PORTRAIT OF **HOW OTHER SAMPLE AVERAGES WOULD PILE UP**...
...**IF THEY WERE CENTERED AT THIS LOCATION!**

NEXT, WE **LOOK BACK** AT THE AVERAGE WE FOUND IN OUR ONE SAMPLE AND ASK:

IF THE POPULATION AVERAGE **WERE REALLY RIGHT HERE...**

...**HOW LIKELY IS IT** THAT WE'D RANDOMLY GRAB A SAMPLE LIKE OURS?

IF OUR GUESS IS CORRECT, AND THE REAL POPULATION AVERAGE **IS RIGHT HERE...**

...THEN **IN THE LONG RUN**, WE'D EXPECT **ANY ONE SAMPLE...**

...TO HAVE AN AVERAGE SOMEWHERE **IN THE CLUMP UNDER THE HUMP.**

HOWEVER, IF OUR ONE SAMPLE HAS AN AVERAGE VALUE **WAY OUT HERE...**

...OR **WAY OUT HERE...**

...WE HAVE REASON TO SUSPECT THAT OUR GUESS **MIGHT BE WRONG.**

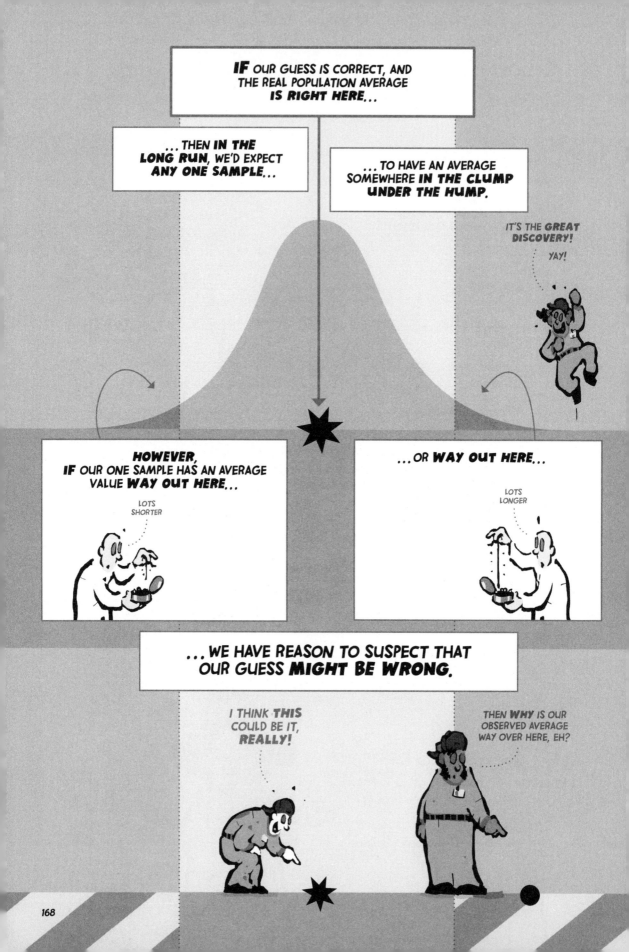

BUT LET'S BE MORE **PRECISE** ABOUT IT:

IN THE LONG RUN, WE EXPECT **95%** OF ALL SAMPLE AVERAGES TO PILE UP **WITHIN TWO STANDARD DEVIATIONS OF THE REAL POPULATION AVERAGE...**

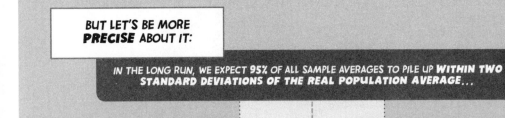

...SO THE **PROBABILITY** THAT WE'D RANDOMLY GRAB A SAMPLE AVERAGE WAY **OUT HERE...**

...OR **OUT HERE...**

...IS ONLY **ABOUT 5%.**

YOU **MIGHT** RANDOMLY GRAB A SAMPLE WAY OUT HERE...

...BUT IT'S **NOT VERY PROBABLE.**

IN PRACTICE, WE COMPARE OUR SAMPLE AND OUR GUESS BY CALCULATING SOMETHING CALLED A **PROBABILITY VALUE*** (OR "P-VALUE")...

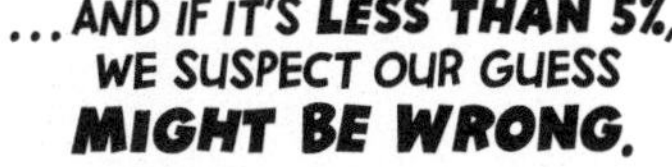

...AND IF IT'S **LESS THAN 5%,** WE SUSPECT OUR GUESS **MIGHT BE WRONG.**

SORRY, THAT'S A LITTLE **TOO UNLIKELY** FOR MY TASTE.

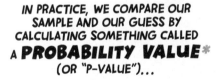

* SEE **PAGES 221–222** FOR A FORMAL DEFINITION.

WE ALWAYS **FINISH** HYPOTHESIS TESTING BY **MAKING A FORMAL DECISION.**

IF OUR SAMPLE AND OUR GUESS ARE **FAIRLY CLOSE TOGETHER...**

WE GOT A P-VALUE OF 5% OR MORE WHEN WE COMPARED THEM...
...WHICH MEANS OUR SAMPLE AVERAGE FITS INSIDE THE 95% CLUMP UNDER THE HUMP.

...WE HAVE TO **CONCLUDE** THAT OUR GUESS **MIGHT BE CORRECT.**

IT'S PRETTY LIKELY THAT WE'D RANDOMLY GRAB A SAMPLE LIKE OURS...
...FROM A POPULATION CENTERED RIGHT THERE.

HOWEVER, IF OUR SAMPLE AND OUR GUESS ARE ***FAR APART***...

WE GOT A P-VALUE OF **LESS THAN 5%** WHEN WE COMPARE THEM...
...WHICH MEANS OUR SAMPLE AVERAGE IS **OUT IN THE TAILS!**

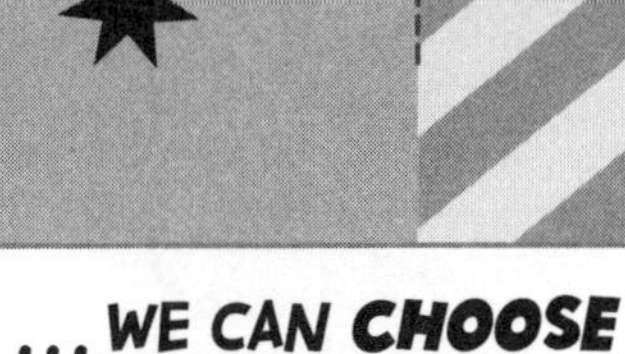

...WE CAN ***CHOOSE TO REJECT IT.***

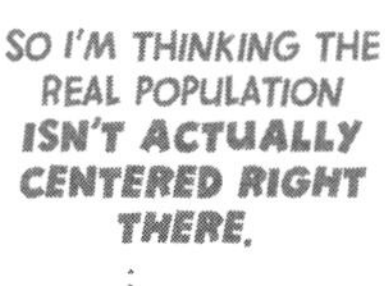

BUT THOSE ARE THE ***ONLY TWO OPTIONS.***

UNFORTUNATELY, **NEITHER** OPTION IS **COMPLETELY SATISFYING.**

BECAUSE **NO MATTER WHICH FORMAL CONCLUSION WE REACH**...

...WE **MIGHT** BE WRONG!*

* *THIS CAN FEEL ANNOYING, BUT IT'S INCREDIBLY IMPORTANT. FOR MORE, SEE PAGE 222.*

REMEMBER, HYPOTHESIS TESTING IS ALL BASED ON **ONE ESTIMATE** BUILT WITH **ONE SAMPLE**...

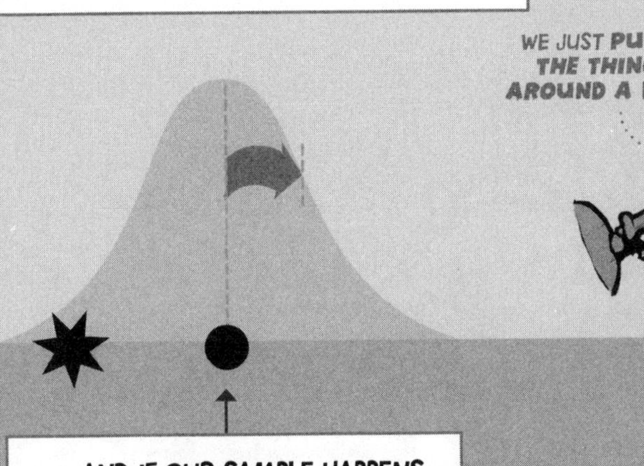

WE JUST **PUSH THE THING AROUND A BIT.**

...AND IF OUR SAMPLE HAPPENS TO BE **MISLEADING**...

...ALL OUR CONCLUSIONS WILL **ALSO BE MISLEADING.**

IF WE RANDOMLY GRABBED **AN ABNORMAL SAMPLE**...

...**OUR P-VALUE** WON'T MEAN SQUAT!

LUCKILY, EVEN THOUGH **COMPLETE CERTAINTY** IS ALWAYS OUT OF REACH...

NO AMOUNT OF **TRAINING** CAN MAKE YOU CORRECT **100%** OF THE TIME.

...THERE **ARE** TIMES WHEN HYPOTHESIS TESTING CAN BE **EXTREMELY USEFUL.**

AS WE'LL SEE IN THE NEXT CHAPTER, THE **TWISTS AND TURNS OF HYPOTHESIS TESTING...**

...WORK BEST WHEN WE'RE WRESTLING WITH ONE **VERY PARTICULAR** KIND OF QUESTION:

IS THIS EVIDENCE **STRONG ENOUGH...**

...TO MAKE US **DOUBT** THIS HYPOTHESIS?

CHAPTER 13
SMACKDOWN

NOW THAT WE'VE LEARNED THE **FORMAL STEPS OF HYPOTHESIS TESTING**...

...LET'S SEE HOW THEY **ACTUALLY GET USED.**

IN PRACTICE, WE USE THESE STEPS...

...**TO PIT ONE IDEA**...

...**AGAINST ANOTHER.**

IT'S LIKE A **WRESTLING MATCH**...

TO SEE HOW IT WORKS, LET'S LOOK AT SOME **STORIES**.

IMAGINE THAT **MINERVA HIGHTOWER** (AKA **DR. HAPPY**)...

HI.

I AM THE WORLD'S FOREMOST PURVEYOR OF **PURE EVIL!**

...IS CONCERNED ABOUT HER **POISON MACHINE.**

IT'S SUPPOSED TO INJECT AN AVERAGE OF **0.25 GRAMS OF PURE EVIL**...

...INTO **EACH VIAL** OF HER PATENTED **SLOW DEATH SALVE**...

BLORP. BLORP.

ALL DR. HAPPY PRODUCTS ARE **PERSONALLY GUARANTEED!**

...BUT IT SEEMS TO BE **ON THE FRITZ**...

BOSS, WE'RE RECEIVING **COMPLAINTS FROM CUSTOMERS**...

...SOME SAY YOUR SALVE **DOESN'T HAVE ENOUGH EVIL IN IT**...

BUT SWEETIE PIE, I'VE GOT A **REPUTATION** TO UPHOLD!

...AND OTHERS SAY IT'S **GOT TOO MUCH EVIL IN IT.**

BUT THAT WOULD BE **WASTEFUL!**

...SO SHE **TESTS IT.**

DR. HAPPY GRABS A **RANDOM SAMPLE**...

...PLOTS IT OUT...

...AND CALCULATES THAT HER **SAMPLE AVERAGE IS 0.14 GRAMS.**

THAT'S **NOT ENOUGH** PURE EVIL IN EVERY VIAL!

THAT OLD MACHINE **MUST BE BROKEN!**

BUT HERE'S THE **PROBLEM**:

DR. HAPPY **REALLY WANTS** TO **BUY A NEW MACHINE**...

...BUT HER ACCOUNTANT WON'T LET HER UNLESS THEY CAN **BE CONFIDENT** THAT THE MACHINE IS **REALLY BROKEN.**

TO **MEASURE HER CONFIDENCE**, DR. HAPPY CAN USE A **HYPOTHESIS TEST.**

DR. HAPPY'S HYPOTHESIS TEST PITS
THESE TWO IDEAS AGAINST EACH OTHER:

ONE IDEA IS **EXCITING...**

...AND ONE IS **DULL.**

EACH IDEA COMES WITH A **DIFFERENT EXPLANATION**
FOR WHY WE GOT THE DATA WE DID...

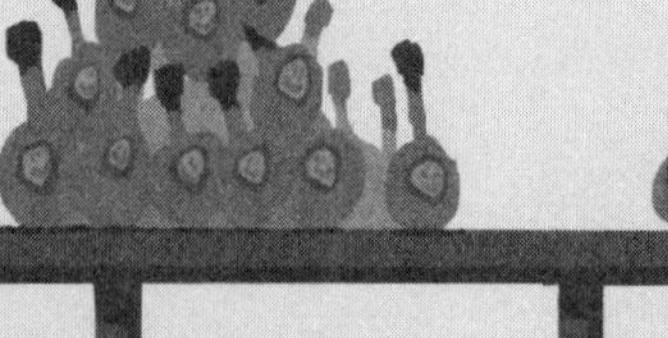

...AND DR. HAPPY'S TASK IS TO LOOK AT THE **DULL** EXPLANATION
AND **SEE IF SHE CAN REJECT IT!**

SO DR. HAPPY USES HER DATA TO **BUILD AN ESTIMATED SAMPLING DISTRIBUTION.**

THEN SHE SLIDES IT TO THE LOCATION PREDICTED BY HER **DULL HYPOTHESIS...**

...AND CALCULATES A **P-VALUE...**

...THEN PAUSES TO THINK ABOUT WHAT HER P-VALUE **ACTUALLY MEANS.**

FINALLY, SHE **CONFIDENTLY REJECTS** THE DULL EXPLANATION...

...IN FAVOR OF **THE EXCITING ONE.**

IN THIS STORY, **DR. HAPPY'S P-VALUE** HELPED HER **MAKE A CONFIDENT DECISION.**

BUT REMEMBER, IN STATISTICS, **CONFIDENCE ALWAYS HAS A FLIP SIDE.**

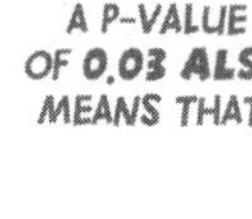

SO EVEN THOUGH THE EVIDENCE **SEEMS TO SUPPORT** DR. HAPPY'S DECISION...

...SHE **COULD** BE WRONG.

FOR BETTER OR FOR WORSE, HYPOTHESIS TESTS **ALWAYS** END THIS WAY.

NEVERTHELESS, IN THE END, DR. HAPPY **GOT WHAT SHE WANTED** FROM HER HYPOTHESIS TEST...

I CAN **DOUBT THE DULL ONE!**
I GET TO BUY THE **XT-4300!**
I HAVE THE **BEST JOB** IN THE WORLD!

...BUT THAT'S **NOT ALWAYS** THE CASE.

TO SEE HOW, LET'S TELL **ANOTHER STORY.**

IMAGINE THAT **CRAZY BILLY** HAS BEEN **POURING WORM STEROIDS** INTO HIS SWAMP...

...AND HE WANTS TO **FIND OUT IF THEY'RE WORKING.**

SO HE GRABS A **RANDOM SAMPLE...**

...PLOTS IT OUT...

THE SAMPLE SIZE IS **30.**

THE SAMPLE AVERAGE IS **4.19** INCHES.

THE SAMPLE SD IS **0.34.**

Frequency

0 1 2 3 4 5 6 7 8

Length (inches)

...AND CALCULATES THAT HIS **SAMPLE AVERAGE IS 4.19 INCHES.**

BEING BILLY, HE **KNOWS** THAT THE SWAMP WORM POPULATION AVERAGE **USED TO BE 4 INCHES...**

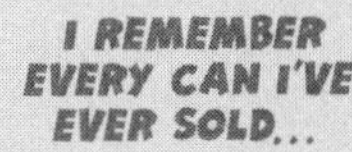

I REMEMBER EVERY CAN I'VE EVER SOLD...

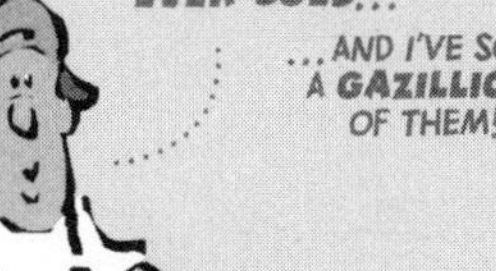

...AND I'VE SOLD A GAZILLION OF THEM!

...AND HE'S **HOPING** TO PROVE **THAT IT HAS GOTTEN LONGER.**

IF IT'S LONGER...
...THESE 'ROIDS ARE WORKING!

IF IT'S NOT LONGER...
...I'VE BEEN HOODWINKED.

HERE'S THE **PROBLEM:**

HE'S GOT A SAMPLE AVERAGE IN HIS HAND THAT'S LONGER **THAN THE OLD AVERAGE...**

THIS SAMPLE MAKES ME THINK MAYBE THE 'ROIDS ARE WORKING!

...BUT IT **MIGHT** BE LONGER **JUST BY CHANCE!**

MAYBE I JUST RANDOMLY GRABBED 30 ABNORMALLY LONG WORMS...
...FROM A POPULATION THAT HASN'T CHANGED!

TO DECIDE WHETHER HE THINKS THAT'S THE CASE, BILLY CAN USE A **HYPOTHESIS TEST.**

CRAZY BILLY'S HYPOTHESIS TEST PITS **THESE TWO IDEAS AGAINST EACH OTHER:**

ONE IDEA IS **NEW AND DIFFERENT...**

...AND ONE IS **RATHER DULL.**

EACH IDEA COMES WITH A **DIFFERENT EXPLANATION** FOR WHY WE GOT THE DATA WE DID...

...AND BILLY'S TASK IS TO LOOK AT THE **DULL** EXPLANATION AND **SEE IF HE CAN REJECT IT!**

SO CRAZY BILLY USES HIS DATA TO **BUILD AN ESTIMATED SAMPLING DISTRIBUTION.**

THEN HE SLIDES IT TO THE LOCATION PREDICTED BY HIS **DULL HYPOTHESIS...**

...AND CALCULATES A **P-VALUE.**

BUT WHEN HE REALIZES WHAT HIS P-VALUE **ACTUALLY MEANS...**

...HE SADLY CONCLUDES THAT HE **CAN'T BE CONFIDENT** THAT THE 'ROIDS ARE WORKING!

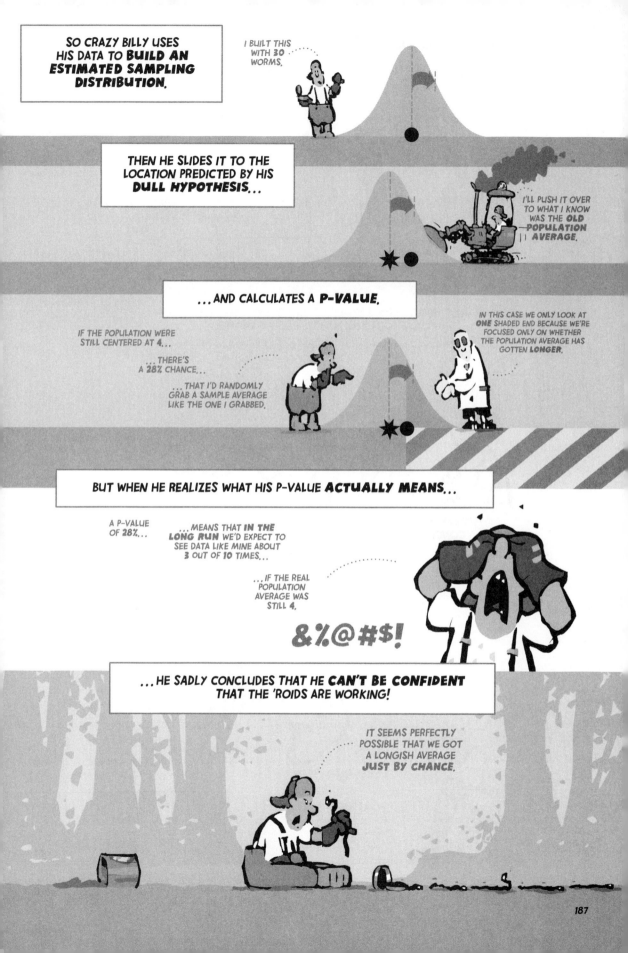

OF COURSE, BILLY'S CONCLUSION...

...DOESN'T MEAN HIS 'ROIDS **AREN'T WORKING.**

IT JUST MEANS **HIS EVIDENCE ISN'T STRONG ENOUGH** TO SUPPORT THE HYPOTHESIS HE WANTED.

IT'S NOT A TERRIBLY **SATISFYING** CONCLUSION TO HIS TEST...

...BUT IT'S AN **IMPORTANT ONE.**

WE'VE NOW SEEN **TWO DIFFERENT HYPOTHESIS TESTS...**

I **WANTED** TO SEE EVIDENCE THAT MY MACHINE WAS BROKEN.

I **WANTED** TO SEE EVIDENCE OF LONGER WORMS.

...WITH **TWO DIFFERENT OUTCOMES.**

BOTH STORIES PITTED **AN ATTRACTIVE NEW IDEA...**

...AGAINST A **TIRED, OLD, DULL ONE.**

IN BOTH STORIES WE **WANTED THE NEW IDEA TO BE TRUE...**

...BUT **GAVE THE DULL IDEA THE BENEFIT OF THE DOUBT.**

THERE'S A **GOOD REASON** WE DO HYPOTHESIS TESTING LIKE THIS.

THE POINT OF HYPOTHESIS TESTING IS TO **MAKE SURE WE DON'T JUMP TO CONCLUSIONS.**

CHAPTER 14
FLYING PIGS, DROOLING ALIENS, AND FIRECRACKERS

LOOKS LIKE STORMY WEATHER.

IN THE LAST SEVERAL CHAPTERS, WE'VE LEARNED HOW TO CALCULATE **CONFIDENCE INTERVALS...**

...AND PERFORM **HYPOTHESIS TESTS.**

AS WE'VE SEEN, **BOTH THESE STRATEGIES** INVOLVE THE **SAME BASIC STEPS.**

...TO GIVE YOU A SENSE OF WHAT LIES AHEAD **IF YOU WANT TO LEARN MORE.**

THE FORECAST CALLS FOR **FLYING PIGS, DROOLING ALIENS,** AND **FIRECRACKERS!**

WE'RE GOING TO WANT **THIS.**

THUS FAR, WE'VE FOCUSED ON HOW OUR **BASIC STEPS** WORK **IN IDEAL CONDITIONS:**

WE'VE LEARNED TO HUNT FOR **ONE POPULATION AVERAGE**...

...USING ONE **LARGE SAMPLE**...

...OF TRULY **RANDOM MEASUREMENTS.**

THE **WATER** IS **CLEAR**...

THE **SUN** IS **SHINING**...

...STATISTICS IS **EASY!**

IN PRACTICE, HOWEVER, CONDITIONS ARE OFTEN **MORE COMPLEX**...

WHAT IF YOU **CAN'T GET A LARGE SAMPLE SIZE?!**

WHAT IF YOU CAN'T GET MEASUREMENTS THAT ARE **REALLY RANDOM?!**

WHAT IF YOU WANT TO KNOW ABOUT **SOMETHING THAT'S NOT AN AVERAGE?**

...AND MUCH OF **ADVANCED STATISTICS** INVOLVES GRAPPLING WITH THESE COMPLEXITIES.

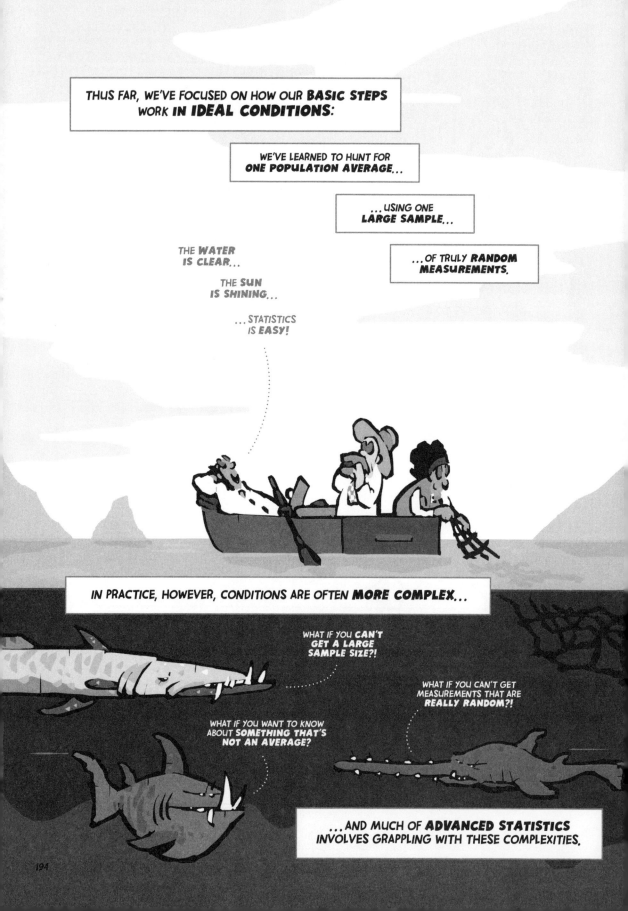

THE GOOD NEWS IS THAT NO MATTER HOW COMPLICATED THINGS GET...

...AND THEY DO GET COMPLICATED...

MY KINGDOM FOR A P-VALUE!

I'M NOT FEELING VERY CONFIDENT!

...WE CAN STILL RELY ON THE **BASIC STEPS** WE'VE COVERED IN THIS BOOK...

...WE JUST HAVE TO LEARN HOW TO **MODIFY** THEM.

THE DETAILS CHANGE...

...BUT THE BASIC STEPS REMAIN THE SAME.

FLYING PIGS!

IN OUR FIRST STORY, LET'S IMAGINE THAT **SPOTTED** FLYING PIGS ARE **FASTER THAN STRIPED** FLYING PIGS...

EAT MY DUST, CUPCAKE!

...BUT THEY'RE ALSO **LOTS** MORE **EXPENSIVE**...

...AND **REVOLTING!**

AND THAT'S WHY SAM NEEDLEHOUSE WANTS TO KNOW: **HOW MUCH FASTER ARE THEY?**

I'M STARTING A **PIG-O-GRAM** BUSINESS.

IS IT BETTER TO **INVEST IN SPOTTED PIGS**...

...OR SHOULD I GET **STRIPED PIGS** INSTEAD, AND USE THE MONEY I SAVE TO **BUY CUTE COSTUMES FOR THEM?**

DO YOU WANNA INVEST IN **SPEED** OR IN **STYLE?**
TO MAKE THAT DECISION, YOU WANT TO HAVE A SENSE OF **HOW MUCH FASTER SPOTTED PIGS ARE.**

IN **STATISTICAL TERMS**, HERE'S THE QUESTION:

HOW DO WE CONSTRUCT A **CONFIDENCE INTERVAL**...
...THAT TELLS US ABOUT THE **DIFFERENCE** BETWEEN **TWO SEPARATE POPULATION AVERAGES?**

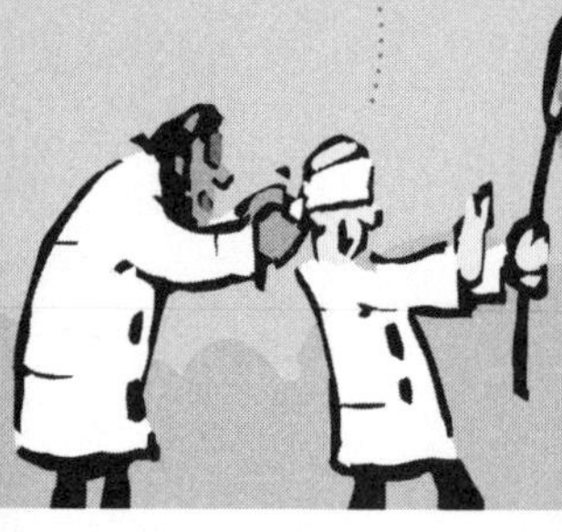

LET'S GRAB **TWO SETS** OF RANDOM SAMPLE DATA TO FIND OUT.

IN THIS CASE, WE CAN USE DATA FROM **40 RANDOM SPOTTED PIGS...**

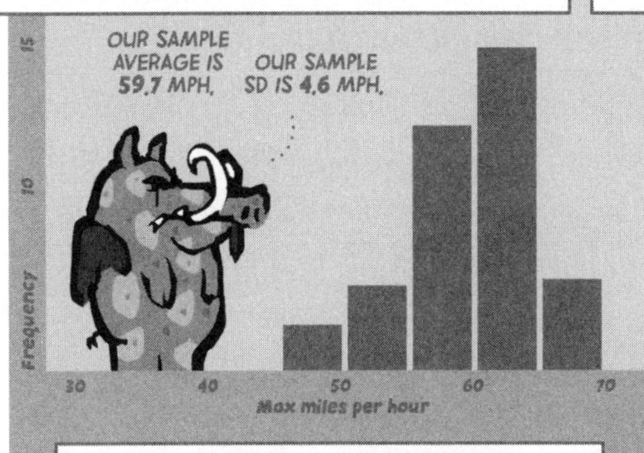

...AND **40 RANDOM STRIPED PIGS...**

...TO BUILD AN **ESTIMATED SAMPLING DISTRIBUTION...**

...THAT'S **SLIGHTLY DIFFERENT** FROM THE ONE WE'RE USED TO.

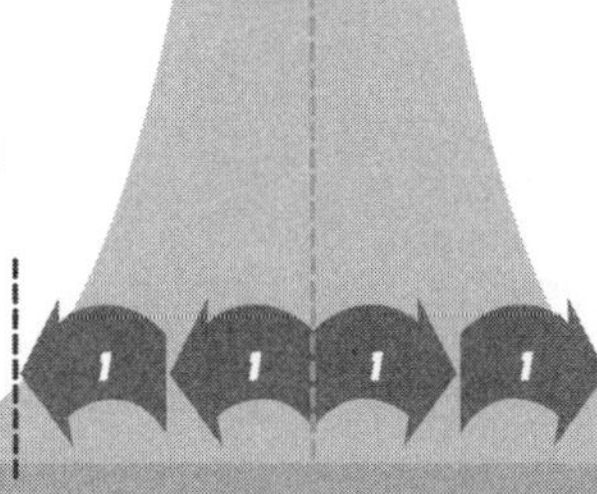

IT'S SLIGHTLY DIFFERENT, BUT WE CAN STILL **CARVE IT UP...***

...AND USE IT TO MAKE A **CONFIDENT DECISION.**

* SEE **PAGE 223** FOR THE FORMULA.

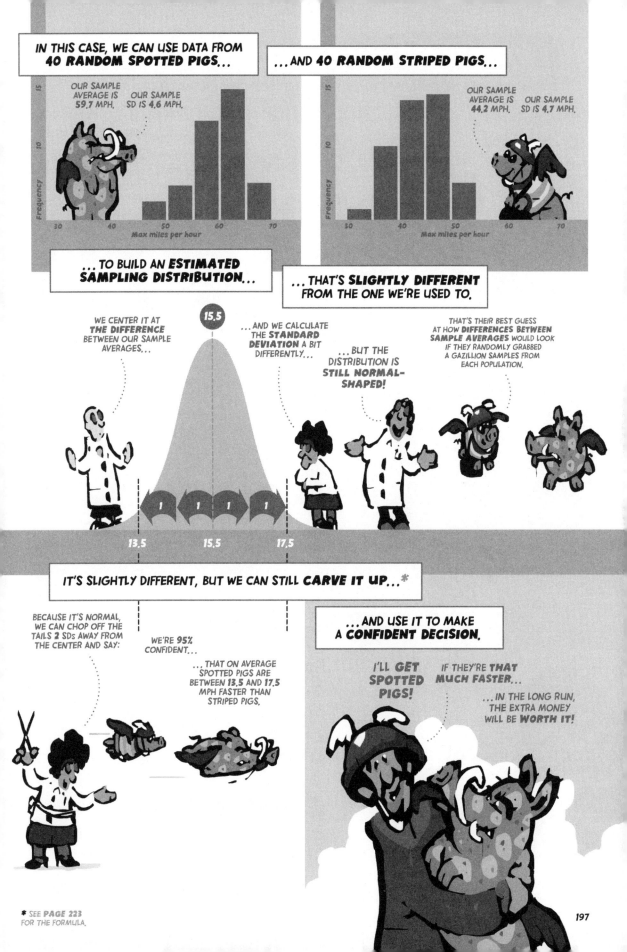

DROOLING ALIENS!

THE **MAN-EATING BUG ALIENS** FROM THE PLANET E8M-286 HAVE **ACIDIC SALIVA**...

...AND WE WANT TO KNOW, **HOW ACIDIC IS IT?**

UNFORTUNATELY, WE CAN GRAB AND MEASURE ONLY **A FEW OF THEM**...

...WHICH **ISN'T ENOUGH** TO MAKE OUR INFERENCE TOOLS WORK THE WAY WE'VE LEARNED SO FAR.

LUCKILY, THERE'S A **CLEVER STRATEGY** WE CAN USE WHEN WE'RE STUCK WITH A **SMALL SAMPLE SIZE**...

...BUT IT BEGINS WITH ONE **WHOPPING ASSUMPTION**.

IF WE'RE WILLING TO **ASSUME** THAT THE **OVERALL POPULATION IS NORMAL-SHAPED...**

SEEMS LIKE A LOT OF **NATURAL PHENOMENA** ARE NORMALLY DISTRIBUTED, SO MAYBE THIS ONE IS, TOO.

...WHICH CAN BE A **DUBIOUS ASSUMPTION...**

IN THIS CASE, IT'S HARD TO TELL WITH JUST **10** ALIENS.

...WE CAN USE OUR SAMPLE DATA TO BUILD AN **ESTIMATED SAMPLING DISTRIBUTION...**

...THAT'S GOT A SLIGHTLY **FATTER SHAPE** THAN THE ONE WE'RE USED TO.

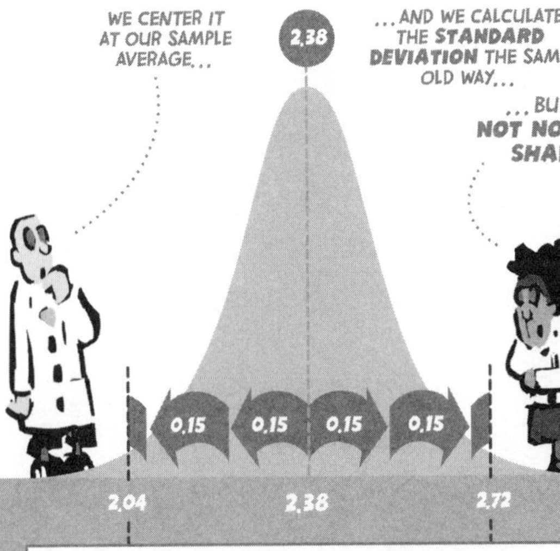

WE CENTER IT AT OUR SAMPLE AVERAGE...
2.38
...AND WE CALCULATE THE **STANDARD DEVIATION** THE SAME OLD WAY...
0.15
0.15
0.15
0.15
2.04
2.38
2.72

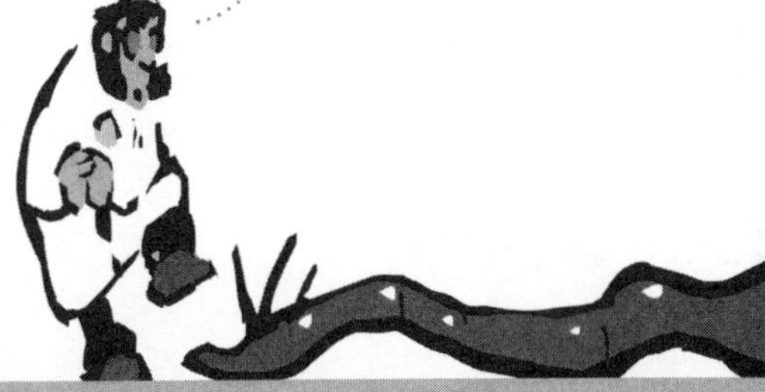

...BUT IT'S **NOT NORMAL-SHAPED!**
IT LOOKS SIMILAR, BUT IT HAS **MORE PROBABILITY OUT IN THE TAILS...**
...IT'S CALLED A **T-DISTRIBUTION.**

IT'S FATTER, BUT WE CAN STILL USE IT TO **CALCULATE OUR CONFIDENCE...***

WHICH MEANS WE'RE **95%** CONFIDENT...
...THAT THEIR AVERAGE SALIVA pH IS BETWEEN **2.04** AND **2.72.**
THAT'S SOMEWHERE BETWEEN **VINEGAR** AND **LEMON JUICE!**

...WE JUST HAVE TO BE **EXTRA CAREFUL ABOUT OUR CONCLUSIONS.**

IF OUR ASSUMPTION ABOUT A NORMAL-SHAPED POPULATION IS **WRONG...**
...**WE MIGHT GET MELTED.**

* SEE **PAGE 223** FOR THE FORMULA.

FIRECRACKERS!

UNFORTUNATELY, LITTLE SUZIE BICKER WANTS TO **HARM THE NEIGHBOR'S CAT...**

... WITH **FIRECRACKERS!**

SHE PREFERS A BRAND CALLED **DINGALINGS...**

... BECAUSE THE PACKAGE CLAIMS THEY HAVE AN **AVERAGE FUSE TIME OF FIVE SECONDS...**

... BUT SHE'S CONCERNED ABOUT **HOW DEPENDABLE THEY ARE.**

IN **STATISTICAL TERMS,** SUZIE'S QUESTION IS ABOUT **VARIABILITY...**

... AND WE CAN TACKLE IT BY USING ONE SAMPLE...

... TO MAKE A GUESS ABOUT THE **SPREAD** OF THE OVERALL POPULATION.

IN THIS CASE, WE WANT TO DO INFERENCE ON A **STANDARD DEVIATION** INSTEAD OF AN AVERAGE...

...AND THE WHOLE PROCESS REQUIRES **VERY DIFFERENT MATH.**

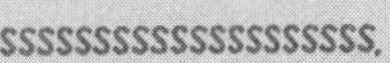

NEVERTHELESS, WE STILL USE THE SAME **BASIC STEPS:**

WE BUILD A SLIGHTLY DIFFERENT KIND OF **ESTIMATED SAMPLING DISTRIBUTION...***

...AND **CARVE PROBABILITIES** OUT OF IT...

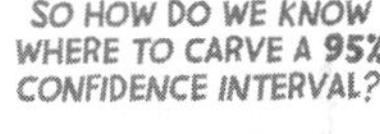

...TO REACH A **FAMILIAR CONCLUSION.**

* SEE **PAGE 224** FOR SOME TECHNICAL DETAILS.

AS THESE STORIES SUGGEST, WE DEPEND ON A **DEEP BAG OF TRICKS...**

...WHEN WE GRAPPLE WITH **ADVANCED STATISTICS QUESTIONS.**

WHAT IF WE WANT TO COMPARE **DROOLING ALIENS** TO **FIRECRACKERS?**
JUST A SEC.

AND IN TRUTH, THE BAG IS **NEARLY BOTTOMLESS.**

FOR EXAMPLE, IF WE HAVE TO CONTEND WITH DATA THAT'S **CORRELATED*** IN SOME WAY...

* SEE **PAGE 224** FOR A MORE TECHNICAL EXPLANATION.

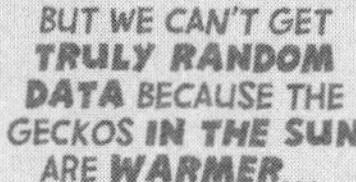
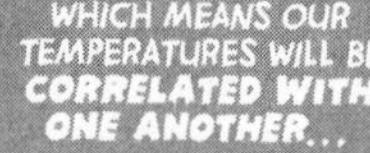

WE WANT TO KNOW THE AVERAGE TEMPERATURE OF ALL THE **GECKOS** IN THIS RAIN FOREST.
BUT WE CAN'T GET **TRULY RANDOM DATA** BECAUSE THE GECKOS **IN THE SUN** ARE **WARMER...**
...THAN THE GECKOS **IN THE SHADE.**
WHICH MEANS OUR TEMPERATURES WILL BE **CORRELATED WITH ONE ANOTHER...**
...AND **SO WILL OURS.**

...WE HAVE **CERTAIN TRICKS** WE CAN USE.

IF WE STRAP A **CORRELATION STRUCTURE** AROUND THE GECKOS...
...WE CAN STILL USE THEM TO **ESTIMATE A SAMPLING DISTRIBUTION!**

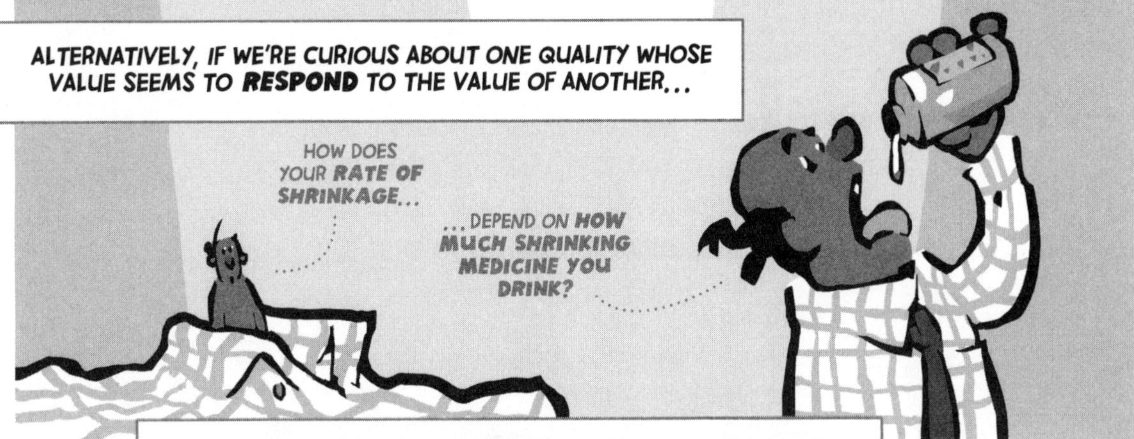

ALTERNATIVELY, IF WE'RE CURIOUS ABOUT ONE QUALITY WHOSE VALUE SEEMS TO **RESPOND** TO THE VALUE OF ANOTHER...
HOW DOES YOUR **RATE OF SHRINKAGE**...
...DEPEND ON **HOW MUCH SHRINKING MEDICINE YOU DRINK?**

...WE HAVE **ENTIRELY DIFFERENT TRICKS** AT OUR DISPOSAL.

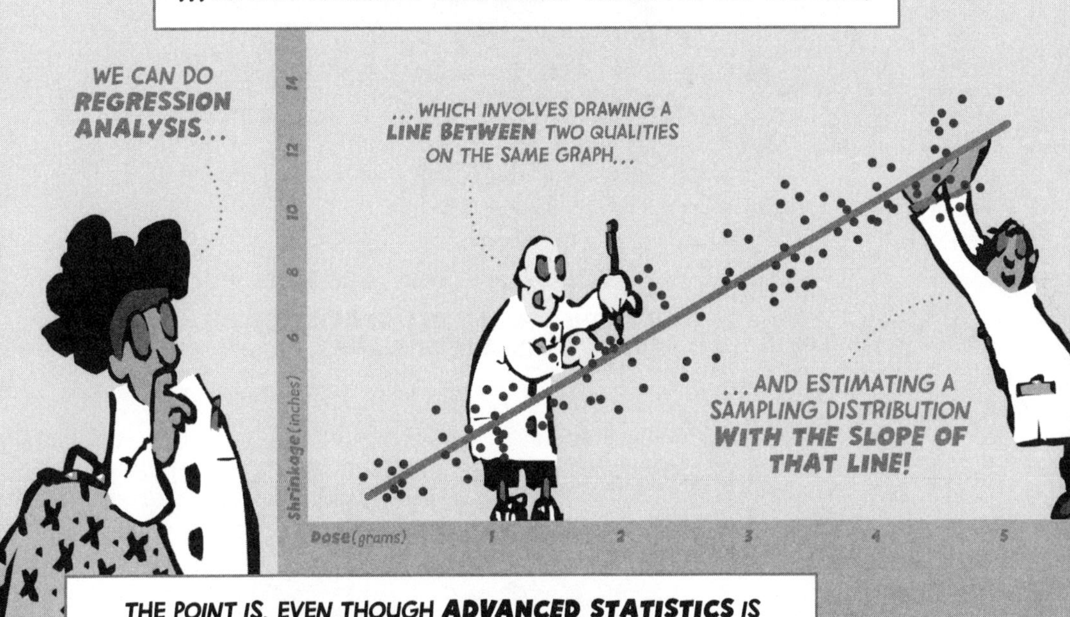

WE CAN DO **REGRESSION ANALYSIS**...
...WHICH INVOLVES DRAWING A **LINE BETWEEN** TWO QUALITIES ON THE SAME GRAPH...
...AND ESTIMATING A SAMPLING DISTRIBUTION **WITH THE SLOPE OF THAT LINE!**
Shrinkage (inches)
6
8
10
12
14
Dose (grams)
1
2
3
4
5

THE POINT IS, EVEN THOUGH **ADVANCED STATISTICS** IS **CRAMMED FULL** OF **TRICKS** AND **TWEAKS**...*

DON'T FORGET **ANOVA**...
...AND HOW TO **DO INFERENCE ON PROPORTIONS**...
...AND HOW TO **PREDICT THE FUTURE!**

...THE **BASIC STEPS** OF STATISTICAL INFERENCE **REMAIN THE SAME!**

* SEE **PAGES 224–225** FOR A MORE THOROUGH RUNDOWN.

SO KEEP THIS IN MIND IF YOU GO ON TO **LEARN MORE STATISTICS**:

THE DETAILS CAN SEEM **OVERWHELMING** AT FIRST...

IF YOU WANT TO KNOW HOW TO **PREDICT THE WEATHER**...
...HERE'S **A WHOLE SEPARATE BAG!**

...BUT AT THEIR HEART, **ALL STATISTICS PROBLEMS ARE SIMILAR.**

THEY **LOOK LIKE THIS**:

HOW DO WE MAKE JUDGMENTS ABOUT **POPULATIONS**...
...WHEN WE ONLY HAVE ACCESS TO **SAMPLES**?

AND WE TACKLE THEM **LIKE THIS:**

WE USE OUR DATA TO ESTIMATE SOME KIND OF **SAMPLING DISTRIBUTION**...
...THEN WE **CARVE PROBABILITIES OUT OF IT**...

...THOUGH IT CAN SOMETIMES BE HELPFUL TO **PUSH IT TO A NEW LOCATION FIRST.**

CONCLUSION

THINKING LIKE A STATISTICIAN

...**GATHERING**...

YOUR **HELMET** IS TO KEEP YOU SAFE.

THIS WILL HELP YOU **AVOID BIAS**.

...AND **HUNTING**.

WE'LL **NEVER** BE ABLE TO ACTUALLY CATCH IT...

...BUT WE CAN MAKE A GUESS ABOUT THE **CLUSTER** THAT SURROUNDS IT.

AND ALL THE WHILE, WE'VE BEEN LEARNING **HOW STATISTICIANS THINK!**

WE DON'T KNOW **EVERYTHING**...

...BUT THAT DOESN'T MEAN WE KNOW **NOTHING!**

IN **PART ONE**, WE **EXAMINED PILES OF SAMPLE DATA**...

TELL US ABOUT YOUR **SHAPE, LOCATION,** AND **SPREAD.**

ARGH!

WE'RE **SKEWED!**

HAR HAR.

...AND **INVESTIGATED THEM.**

BEWARE **LURKING VARIABLES!**

THEN, IN **PART TWO,** WE STUDIED **STATISTICAL INFERENCE**...

... WHICH IS HOW WE USE **SAMPLES** TO **SEARCH FOR QUALITIES IN AN OVERALL POPULATION.**

MORE SPECIFICALLY, WE LEARNED HOW TO BUILD **ESTIMATED SAMPLING DISTRIBUTIONS...**

...AND HOW TO **PEER** INTO THEM...

...AND **CARVE THEM UP** TO **CALCULATE CONFIDENCE INTERVALS...**

...OR **PUSH THEM AROUND** TO PERFORM **HYPOTHESIS TESTS.**

FINALLY, WE LEARNED HOW WE CAN **MODIFY** THESE BASIC STEPS...

POW!

...WHEN OUR QUESTIONS GET **MORE COMPLICATED.**

SINCE THE **BASIC STEPS** OF STATISTICAL INFERENCE WERE FIRST DISCOVERED...

...THEY'VE BEEN **ADAPTED** BY ALL KINDS OF DIFFERENT PEOPLE...

...FOR USE IN ALL KINDS OF **DIFFERENT CIRCUMSTANCES.**

WHILE THIS IS A TESTAMENT TO THE INCREDIBLE **POWER** OF STATISTICS...

...IT HAS CREATED A BIT OF A **CONUNDRUM.**

OVER THE YEARS, STATISTICAL TERMINOLOGY HAS FLOURISHED...
...AND MULTIPLIED...
THE NORMAL DISTRIBUTION...
...IS ALSO CALLED THE GAUSSIAN DISTRIBUTION...
...IS ALSO CALLED THE Z-DISTRIBUTION.
...UM.
...AND AS A RESULT, LEARNING HOW TO TALK LIKE A STATISTICIAN...
...CAN BE ROUGH...
OUCH, I THINK MY EVENTS MIGHT BE DISJOINT!
DO I WANT SIGNIFICANCE OR POWER?
IS MY DEGREE OF FREEDOM RESIDUAL?
IS THIS ERROR STANDARD?
...ESPECIALLY IF YOU MOVE ON TO MASTER THE MORE ADVANCED TOOLS.
REPEAT AFTER ME:
P-VALUE,
E-VALUE,
Z- OR T-SCORES,
CHI-SQUARED G-TESTS,
AND THERE'S EVEN MORE!
I'M NOT FEELING VERY CONFIDENT AT THE MOMENT.
HAVE YOU CONSIDERED A NON-PARAMETRIC REGRESSION SIMULATION?

KEEP **ONE EYE ON THE LONG RUN...**

...AND **ONE EYE ON THE SHORT RUN...**

...AT THE SAME TIME.

ANYONE CAN DO IT!

BUT IF YOU WANT TO LEARN **HOW THEY TALK...**

...YOU CAN **START BY EXPLORING...**

APPENDIX

THE MATH CAVE

ABANDON HOPE ALL YE WHO ENTER HERE...

...UNLESS YOU WANT TO LEARN TO TALK LIKE A STATISTICIAN...

...AND WRITE LIKE ONE, TOO!

WHEN WE WRITE FORMULAS WE LIST ALL THE OBSERVATIONS IN OUR SAMPLE LIKE THIS:

$$x_1, x_2, x_3 \ldots x_n$$

SO THAT x_1 IS THE 1st OBSERVATION...

...x_2 IS THE 2nd OBSERVATION...

...AND x_n IS OUR FINAL OBSERVATION IN A LIST THAT HAS **n** OBSERVATIONS IN IT.

FROM PP. 36-37

RANDOM SAMPLING

RANDOM SAMPLING IS ABSOLUTELY ESSENTIAL TO STATISTICAL INQUIRY. THE KEY FEATURE OF A RANDOM SAMPLE IS THAT IT **DOES NOT DIFFER SYSTEMATICALLY** FROM THE POPULATION IT COMES FROM.

TECHNICALLY, A **SAMPLE** IS A COLLECTION OF SEPARATE OBSERVATIONS ABOUT A SPECIFIC **VARIABLE** (SEE BELOW). WE CALL IT A **RANDOM SAMPLE** WHEN IT'S MADE UP OF RANDOMLY GATHERED OBSERVATIONS, EACH OF WHICH IS **INDEPENDENT** OF ALL THE OTHERS.

WHEN WE TALK IN THIS BOOK ABOUT RANDOM SAMPLING, WE SPECIFICALLY MEAN **SIMPLE RANDOM SAMPLING**. FORMALLY, A **SIMPLE RANDOM SAMPLE (SRS)** OF SIZE **n** IS A COLLECTION OF **n** OBSERVATIONS OBTAINED IN SUCH A WAY THAT **ALL POSSIBLE SAMPLES** OF **n** OBSERVATIONS FROM THE POPULATION ARE **EQUALLY LIKELY TO HAVE BEEN SELECTED.**

SOME OTHER **NON-RANDOM SAMPLING TECHNIQUES** SUCH AS SYSTEMATIC SAMPLING AND STRATIFIED SAMPLING SOMETIMES ALSO WORK. BUT WHATEVER SAMPLING STRATEGY WE END UP USING, WE MUST BE CERTAIN THAT THE RESULTING SAMPLE IS **REPRESENTATIVE OF THE POPULATION**. IF IT'S NOT, EVERYTHING THAT FOLLOWS IS WORTHLESS.

FROM P. 54

SAMPLE SIZE (n)

THE SAMPLE SIZE IS THE TOTAL NUMBER OF MEASUREMENTS INCLUDED IN A SINGLE SAMPLE. IN GENERAL, A LARGER **n** INCREASES THE **CONFIDENCE** WE CAN HAVE IN OUR STATISTICAL CONCLUSIONS, BUT **ONLY** IF OUR SAMPLE IS **RANDOM!**

FROM P. 59

SAMPLE AVERAGE ($\bar{x}$)

WE COMPUTE THE **AVERAGE** IN A SAMPLE BY ADDING UP ALL VALUES IN THAT SAMPLE AND DIVIDING BY THE SAMPLE SIZE. HERE'S THE FORMULA:

ARRRRR, WE CALL OUR SAMPLE AVERAGE "**XBAR.**"

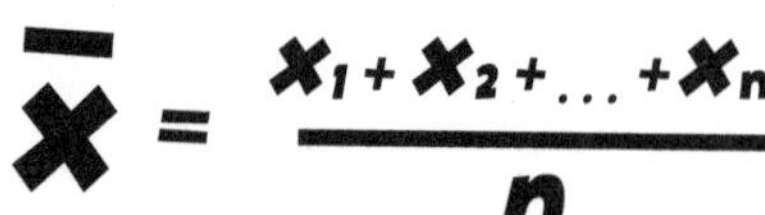

$$\bar{x} = \frac{x_1 + x_2 + \ldots + x_n}{n}$$

THE AVERAGE IS ALSO COMMONLY KNOWN AS THE "**ARITHMETIC MEAN**," OR JUST "**THE MEAN**" FOR SHORT. IN THIS BOOK WE'VE AVOIDED "MEAN" AND USED "AVERAGE" INSTEAD BECAUSE WE HOPE THAT BY USING THIS MORE FAMILIAR TERM WE CAN HELP MAKE STATISTICAL INFERENCE FEEL MORE FAMILIAR. ALSO, WE BELIEVE MOST READERS THINK OF THE ARITHMETIC MEAN WHEN THEY HEAR THE WORD "AVERAGE" ANYWAY.

WHATEVER YOU CALL IT, THE AVERAGE IS THE MOST BASIC MEASURE OF THE **CENTRAL TENDENCY** IN A DISTRIBUTION. THERE ARE SEVERAL OTHER WAYS TO REFINE OUR UNDERSTANDING OF HOW A PARTICULAR DATA SET CLUMPS TOGETHER, BUT THE CHOICE OF WHICH TO USE DEPENDS ON THE SITUATION.

FOR EXAMPLE, THE **MEDIAN** IS THE "MIDDLE VALUE" OF A SAMPLE AND MAY BE PREFERABLE IN CASES OF SKEW. SIMILARLY, A **TRIMMED AVERAGE** IS COMPUTED BY EXCLUDING A SMALL PERCENTAGE OF THE SMALLEST AND LARGEST VALUES, AND MAY BE PREFERABLE WHEN THERE ARE EXTREME VALUES IN A SAMPLE.

FROM P. 65

STANDARD DEVIATION (s)

OUR GOAL WHEN WE CALCULATE **STANDARD DEVIATION** IS TO GET A SENSE OF THE AVERAGE DISTANCE FROM THE AVERAGE VALUE. HERE'S HOW TO DO IT IN (MOSTLY) PLAIN ENGLISH:

1) CALCULATE THE DISTANCE BETWEEN EACH MEASUREMENT x AND THE SAMPLE AVERAGE $\bar{x}$. WE CALL THIS DISTANCE A **DEVIATION**.

2) SQUARE EACH DEVIATION.

3) ADD UP ALL THE SQUARED DEVIATIONS.

4) DIVIDE THE SUM BY $n - 1$ (IF WE STOP HERE, WE GET WHAT'S CALLED THE **VARIANCE**.)

5) TAKE THE SQUARE ROOT OF THE WHOLE SHEBANG.

HERE'S THE ACTUAL FORMULA:

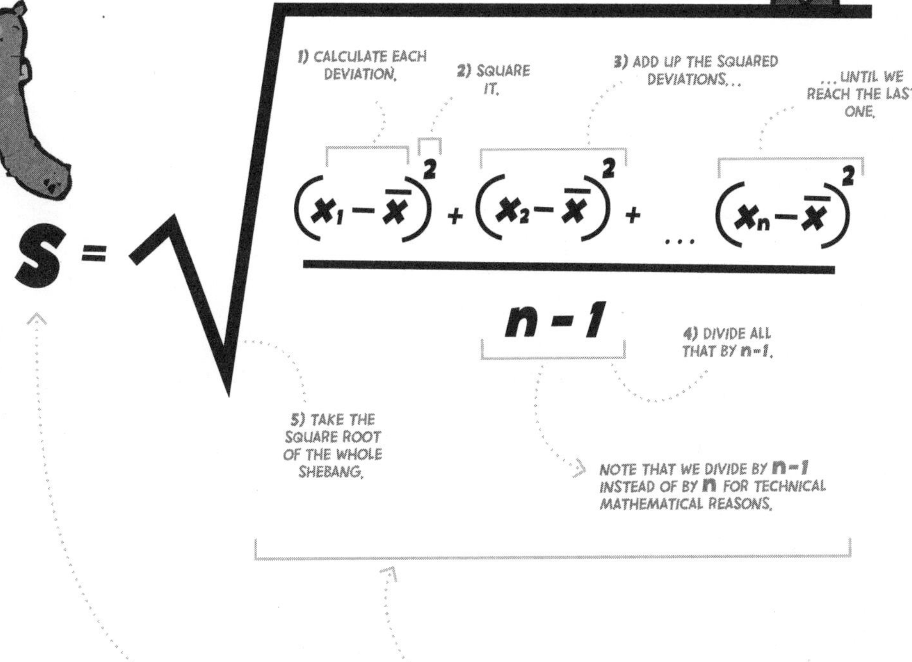
1) CALCULATE EACH DEVIATION.
2) SQUARE IT.
3) ADD UP THE SQUARED DEVIATIONS...
...UNTIL WE REACH THE LAST ONE.
4) DIVIDE ALL THAT BY n-1.
5) TAKE THE SQUARE ROOT OF THE WHOLE SHEBANG.
NOTE THAT WE DIVIDE BY n-1 INSTEAD OF BY n FOR TECHNICAL MATHEMATICAL REASONS.

$$s = \sqrt{\frac{(x_1 - \bar{x})^2 + (x_2 - \bar{x})^2 + \dots (x_n - \bar{x})^2}{n - 1}}$$

TECHNICALLY, THE **VARIANCE** IS THE AVERAGE OF THE SQUARED DIFFERENCES FROM THE AVERAGE, AND THE **STANDARD DEVIATION** IS THE SQUARE ROOT OF THE VARIANCE. NOTE THAT WE USE THE SINGLE LETTER **S** TO REFER SPECIFICALLY TO THE STANDARD DEVIATION OF OUR SAMPLE.

FROM P. 70

VARIABLE (X)

A VARIABLE IS A PARTICULAR QUALITY WE'RE CURIOUS ABOUT. HOWEVER, BECAUSE IN STATISTICS WE ALWAYS COLLECT DATA RANDOMLY, WE REFER TO THE VARIABLES WE'RE LOOKING AT AS **RANDOM VARIABLES**. TECHNICALLY, A **RANDOM VARIABLE** IS A VARIABLE WHOSE VALUE IS RANDOM.

IN THE SHORT TERM, WE HAVE NO WAY OF PREDICTING A RANDOM VARIABLE'S VALUE BEFORE WE GATHER IT. IT'S LIKE A COIN FLIP. IN THE LONG TERM, WE PREDICT THE VALUE OF A RANDOM VARIABLE USING **PROBABILITY** (SEE BELOW).

WORM LENGTH IS A VARIABLE.

SO IS **PIRATE INCOME**.

SO IS **DRAGON SPEED**.

FROM P. 84

DISTRIBUTIONS

IN GENERAL MATHEMATICAL TERMS, THE WORD **DISTRIBUTION** DESCRIBES THE ARRANGEMENT OF ALL THE POSSIBLE VALUES FOR A RANDOM VARIABLE. IF, FOR EXAMPLE, YOU MADE A HISTOGRAM OF ALL THE VALUES OF A VARIABLE IN AN ENTIRE POPULATION, YOU'D BE LOOKING AT THE **POPULATION DISTRIBUTION** FOR THAT VARIABLE.

MORE GENERALLY, DISTRIBUTIONS ALLOW US TO COMPUTE **PROBABILITIES** (OR LONG-RUN LIKELIHOODS) OF RANDOMLY GRABBING VALUES FROM PARTICULAR INTERVALS. IN STATISTICAL INFERENCE, WE CALCULATE PROBABILITIES USING **SAMPLING DISTRIBUTIONS** (SEE BELOW), BUT IF WE HAD A POPULATION DISTRIBUTION IN FRONT OF US, WE COULD ALSO USE IT TO CALCULATE PROBABILITIES. HERE'S HOW:

IF WE SOMEHOW KNEW HOW THE **ENTIRE POPULATION** OF FISH IN A LAKE, SORTED BY LENGTH, WAS DISTRIBUTED...

...WE COULD DO SOME MATH TO CALCULATE THE PROPORTION OF FISH INSIDE ANY AREA OF THAT DISTRIBUTION...

...LIKE **THIS AREA** COVERING THE RANGE FROM 8 TO 12 INCHES.

IT FOLLOWS THAT IF WE REACHED INTO THE LAKE AND **RANDOMLY GRABBED ONE FISH**...

...THE **PROBABILITY** THAT IT WOULD HAVE A LENGTH BETWEEN 8 AND 12 INCHES IS THE SAME AS THE PROPORTION OF THE TOTAL DISTRIBUTION THAT'S INSIDE THE DARKER AREA.

IF **HALF** OF ALL THE FISH ARE BETWEEN 8 AND 12 INCHES...

...THE PROBABILITY I'LL RANDOMLY CATCH ONE IN THAT RANGE IS 50%, OR **0.5**.

0 1 2 3 4 5 6 7 8 9 10 11 12 13 14 15 16

OF COURSE, IN REALITY, WE NEVER ACTUALLY GET TO LOOK AT AN ENTIRE POPULATION DISTRIBUTION. IF WE DID, WE WOULDN'T NEED STATISTICS.

FROM P. 85

SAMPLE STATISTICS VS. POPULATION PARAMETERS

SINCE OUR GOAL IN STATISTICS IS ALWAYS TO USE **SAMPLES** TO MAKE GUESSES ABOUT **POPULATIONS**, WE HAVE DIFFERENT TERMS AND TECHNICAL NOTATION FOR EACH.

WE CALL QUALITIES IN A SAMPLE "**STATISTICS**."

WHEN WE'RE WRITING FORMULAS, **XBAR** REFERS EXCLUSIVELY TO OUR **SAMPLE AVERAGE:** $\bar{x}$

S REFERS EXCLUSIVELY TO OUR **SAMPLE STANDARD DEVIATION:**

STATISTICS ARE THE THINGS WE ACTUALLY MEASURE AND THEREFORE KNOW WITH CERTAINTY.

WE CALL QUALITIES IN A POPULATION "**PARAMETERS**."

THE LOWERCASE GREEK LETTER **MU** REFERS EXCLUSIVELY TO THE **POPULATION AVERAGE:** μ

THE LOWERCASE GREEK LETTER **SIGMA** REFERS EXCLUSIVELY TO THE **POPULATION STANDARD DEVIATION:** σ

PARAMETERS ARE THE THINGS WE REALLY WANT TO KNOW, BUT CAN ONLY MAKE GUESSES ABOUT.

FROM P. 94

THE NORMAL DISTRIBUTION

IN MATHEMATICS AND PROBABILITY THEORY, THERE ARE LOTS OF DIFFERENT KINDS OF DISTRIBUTIONS THAT COME IN LOTS OF DIFFERENT SHAPES. BY FAR THE MOST FAMOUS HOWEVER, IS THE **NORMAL DISTRIBUTION.** IN STATISTICS, WE CARE MOST ABOUT IT BECAUSE IT'S HOW AVERAGES TEND TO PILE UP (SEE THE **CLT**, BELOW).

LIKE ANY OTHER DISTRIBUTION, WE CAN CARVE UP A NORMAL DISTRIBUTION INTO AREAS THAT DEPICT PROBABILITIES FOR THE VALUES INSIDE IT. WE LEARN HOW TO DO THIS ON PAGE **115**, BUT HERE'S AN EXAMPLE:

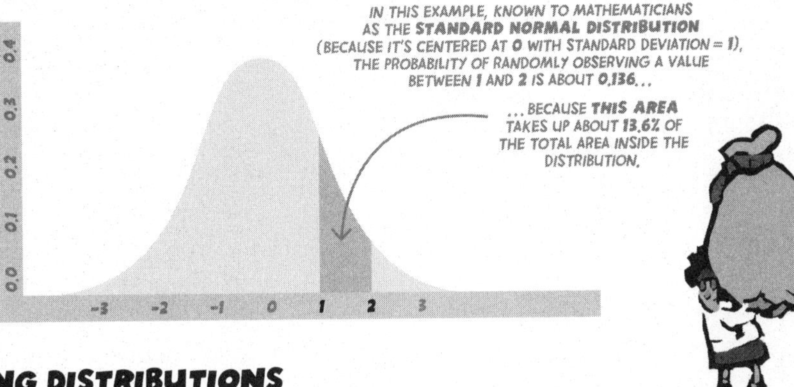
IN THIS EXAMPLE, KNOWN TO MATHEMATICIANS AS THE **STANDARD NORMAL DISTRIBUTION** (BECAUSE IT'S CENTERED AT 0 WITH STANDARD DEVIATION = 1), THE PROBABILITY OF RANDOMLY OBSERVING A VALUE BETWEEN 1 AND 2 IS ABOUT 0.136...
...BECAUSE **THIS AREA** TAKES UP ABOUT 13.6% OF THE TOTAL AREA INSIDE THE DISTRIBUTION.
0.0
0.1
0.2
0.3
0.4
-3
-2
-1
0
1
2
3

FROM P. 95

SAMPLING DISTRIBUTIONS

TECHNICALLY, A **SAMPLING DISTRIBUTION** IS THE DISTRIBUTION OF A SAMPLE STATISTIC. ALTHOUGH WE CAN BUILD SAMPLING DISTRIBUTIONS FOR ANY STATISTIC (STANDARD DEVIATIONS, MEDIANS, ETC.) WE'RE FOCUSING HERE ON SAMPLING DISTRIBUTIONS MADE OF AVERAGES. SO, FOR EXAMPLE, IF WE COLLECTED MANY, MANY SAMPLES OF SIZE n FROM A POPULATION, COMPUTED $\bar{x}$ FOR EACH, THEN MADE A HISTOGRAM OF ALL THE $\bar{x}$ VALUES, WE'D BE LOOKING AT THE SAMPLING DISTRIBUTION OF $\bar{x}$. THE PILE IN CRAZY BILLY'S BAIT BARN IS AN EXAMPLE (SEE PAGE **107**). SAMPLING DISTRIBUTIONS ARE **KEY** TO STATISTICAL INFERENCE.

FROM P. 101–102

THE CENTRAL LIMIT THEOREM (CLT)

MUCH OF STATISTICAL INFERENCE DEPENDS ON THE **CENTRAL LIMIT THEOREM**, WHICH STATES THAT THE SAMPLING DISTRIBUTION OF $\bar{x}$ BECOMES APPROXIMATELY **NORMAL** AS THE SAMPLE SIZE n GETS LARGE.

MORE SPECIFICALLY, FOR **RANDOM SAMPLES** OF LARGE SIZE n TAKEN FROM A SINGLE POPULATION WITH AVERAGE μ AND SD σ, THE DISTRIBUTION OF $\bar{x}$ IS APPROXIMATELY **NORMAL** WITH AVERAGE μ AND SD EQUAL TO $\sigma/\sqrt{n}$.

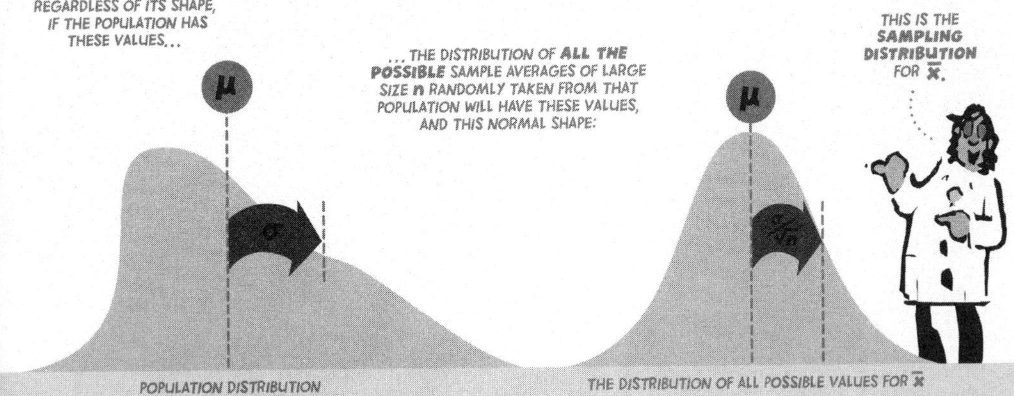
REGARDLESS OF ITS SHAPE, IF THE POPULATION HAS THESE VALUES...
...THE DISTRIBUTION OF **ALL THE POSSIBLE** SAMPLE AVERAGES OF LARGE SIZE n RANDOMLY TAKEN FROM THAT POPULATION WILL HAVE THESE VALUES, AND THIS NORMAL SHAPE:
THIS IS THE **SAMPLING DISTRIBUTION** FOR $\bar{x}$.
μ
σ
μ
$\sigma/\sqrt{n}$
POPULATION DISTRIBUTION
THE DISTRIBUTION OF ALL POSSIBLE VALUES FOR $\bar{x}$
$\sigma/\sqrt{n}$ IS ALSO KNOWN AS THE **STANDARD ERROR.**

THE CENTRAL LIMIT THEOREM (CONT.)

THE CLT IS A VERY GENERAL RESULT THAT WILL ALMOST ALWAYS APPLY AS DESCRIBED IN THE BOOK. THAT SAID, THERE ARE **IMPORTANT CONDITIONS** UNDERLYING THE CLT.

FIRST, THE CLT ONLY WORKS IF EACH OF THE VALUES FOR $X_1, X_2, X_3 \ldots X_n$ IN OUR SAMPLE COMES FROM THE **SAME EXACT POPULATION DISTRIBUTION**. THIS WILL USUALLY BE TRUE FOR SAMPLES OBTAINED IN PRACTICE, BUT CAN BE RELEVANT IF WE'RE INVESTIGATING MORE COMPLICATED QUESTIONS.

SECOND, EACH MEASUREMENT X_i HAS TO BE **RANDOM**. TECHNICALLY THIS ALSO MEANS THAT ALL VALUES FOR X_i HAVE TO BE **INDEPENDENT** OF ONE ANOTHER, SO THAT THE VALUE OF EACH MEASUREMENT X_i DOES NOT DEPEND ON THE VALUES OF THE OTHER SAMPLE VALUES. FOR EXAMPLE, MEASUREMENTS OF TEMPERATURE TAKEN ACROSS A GEOGRAPHICAL REGION WILL **NOT** BE INDEPENDENT, SINCE THE TEMPERATURE AT ONE LOCATION WILL TEND TO BE SIMILAR TO THE TEMPERATURE AT A NEARBY LOCATION; STATISTICIANS WOULD SAY THESE MEASUREMENTS ARE "**CORRELATED**," BECAUSE THERE EXISTS A SYSTEMATIC UNDERLYING PATTERN THAT INFLUENCES THE VALUE OF EACH X_i (SEE **CORRELATION**, BELOW.)

FINALLY, AND MOST TECHNICALLY, THE CENTRAL LIMIT THEOREM APPLIES WHEN n APPROACHES INFINITY, BUT FOR PRACTICAL PURPOSES WE USE AN **APPROXIMATE VERSION** OF THE CLT THAT WORKS WHEN $n \geq 30$. AS A RESULT, IN PRACTICE WE CONSIDER ANY SAMPLE SIZE $n \geq 30$ TO BE "**LARGE**." THIS OFTEN FEELS ARBITRARY, BUT A MORE THOROUGH EXPLANATION WOULD REQUIRE LOTS MORE MATH.

FROM P. 112

PROBABILITIES

IN THE BOOK WE NOTE PROBABILITIES AS PERCENTAGES (E.G., **95%**), BUT IN MATHEMATICS WE USE NUMBERS BETWEEN **0** AND **1** TO EXPRESS THE SAME THING (E.G., **95% = 0.95**). SO, FORMALLY, A **PROBABILITY** IS A NUMBER BETWEEN **0** AND **1** THAT QUANTIFIES THE LIKELIHOOD THAT A RANDOM EVENT WILL OCCUR; THE CLOSER THE PROBABILITY IS TO **1** (OR 100%), THE MORE LIKELY THE EVENT IS TO OCCUR, IN THE LONG RUN. IN OTHER WORDS, PROBABILITIES ARE LIKE PREDICTIONS ABOUT THE LONG RUN. THE TRICKY THING ABOUT THEM, HOWEVER, IS THAT THEY **ONLY REFER TO THE LONG RUN**.

IF, FOR EXAMPLE, THERE ARE EQUAL NUMBERS OF MALE AND FEMALE VOTERS IN A STATE, THE PROBABILITY THAT A RANDOMLY SELECTED VOTER IS FEMALE IS **0.5**. HOWEVER, THE FIRST FEW VOTERS RANDOMLY SAMPLED MAY WELL BE ALL MALE, JUST BY CHANCE. THE **0.5** SPEAKS TO WHAT WOULD HAPPEN **IN THE LONG RUN**: IF WE RANDOMLY SAMPLE ENOUGH VOTERS, WE WILL **EVENTUALLY** END UP WITH **ROUGHLY EQUAL NUMBERS** OF MALE AND FEMALE VOTERS.

IN ANOTHER EXAMPLE, WHEN WE FLIP A COIN, THERE'S A PROBABILITY OF **0.5** THAT IT WILL LAND ON HEADS. BUT EVEN IF WE JUST FLIPPED THE COIN ONCE AND GOT HEADS, THE PROBABILITY OF THE NEXT FLIP LANDING ON HEADS IS **STILL 0.5**. IN THIS WAY, EACH FLIP IS **INDEPENDENT** OF THE OTHERS.

IN SUM, ANY TIME WE CALCULATE A PROBABILITY, IT CAN BE EXPRESSED AS A NUMBER BETWEEN **0** AND **1** (OR, EQUIVALENTLY, **0** AND **100%**), AND THAT NUMBER ALWAYS CORRESPONDS TO THE AREA INSIDE A PROBABILITY DISTRIBUTION. BY DEFINITION, THE TOTAL AREA INSIDE ANY PROBABILITY DISTRIBUTION EQUALS **1**.

FROM P. 114

PROBABILITY MATH

TECHNICALLY, WE CAN COMPUTE AREAS INSIDE ANY DISTRIBUTION (LIKE THE NORMAL ONE DEPICTED ON PAGE 114) USING **INTEGRATION**, WHICH IS A **CALCULUS TECHNIQUE**. IN PRACTICE, STATISTICIANS ASK COMPUTERS TO DO THE CALCULATIONS FOR THEM.

IN THE BILLY'S BAIT BARN EXAMPLE, THE SAMPLING DISTRIBUTION IS NORMAL-SHAPED BECAUSE OF THE CENTRAL LIMIT THEOREM. HOWEVER, IN A LOT OF OTHER STATISTICAL APPLICATIONS, A PARTICULAR SAMPLING DISTRIBUTION WON'T BE NORMAL-SHAPED, BUT WE CAN STILL DO CALCULATIONS LIKE THESE, USING CALCULUS. FOR MORE ABOUT THAT, SEE CHAPTER 14.

ALL DISTRIBUTIONS CAN BE DRAWN AS CURVES, BUT THEY CAN ALSO BE WRITTEN AS **FUNCTIONS**, WHICH ARE LIKE MATH MACHINES THAT TAKE **INPUTS** (IN THIS CASE A RANDOM VARIABLE) AND TURN THEM INTO **OUTPUTS** (IN THIS CASE A PROBABILITY).

IN **MATH NOTATION**, HERE'S A GENERIC WAY TO WRITE ABOUT A PROBABILITY FUNCTION f WITH AVERAGE μ AND STANDARD DEVIATION σ:

IF X IS A **DISCRETE RANDOM VARIABLE** WITH DISTRIBUTION $f_{\mu,\sigma}$...

... THEN $f_{\mu,\sigma}(x)$ EQUALS THE PROBABILITY THAT X TAKES THE VALUE x.

UNFORTUNATELY, IT GETS EVEN MORE COMPLICATED FAST. FOR EXAMPLE, HERE'S THE PROBABILITY FUNCTION FOR THE NORMAL DISTRIBUTION:

$$h_{\mu,\sigma}(x) = \frac{1}{\sigma\sqrt{2\pi}} \exp\left\{-\frac{1}{2\sigma^2}(x-\mu)^2\right\}$$

THOUGH THIS NOTATION IS **TERRIFYING** AT FIRST GLANCE, IN THE SCOPE OF BROADER STATISTICAL AND MATHEMATICAL INQUIRY, PROBABILITY FUNCTIONS LIKE THIS ARE ENORMOUSLY USEFUL BECAUSE THEY RELATE **PARTICULAR KINDS OF RANDOM EVENTS** (LIKE CATCHING A CERTAIN SIZE FISH) WITH **PREDICTABLE LONG-RUN OUTCOMES** (HOW OFTEN YOU'D EXPECT THAT TO HAPPEN IN THE LONG RUN).

FROM P. 129

ESTIMATING A SAMPLING DISTRIBUTION

IN PRACTICE, WHEN WE MAKE USE OF THE CLT, WE HAVE NO WAY OF KNOWING THE REAL VALUES FOR THE PARAMETERS μ AND σ, SO WE USE THE STATISTICS $\bar{x}$ AND s TO **APPROXIMATE** THEM. THIS APPROXIMATION WORKS BECAUSE WE GATHER OUR STATISTICS RANDOMLY. AS A RESULT, WE EXPECT $\bar{x}$ TO DIFFER FROM μ AND s TO DIFFER FROM σ, BUT **ONLY BECAUSE OF CHANCE VARIATION**.

AFTER WE'VE SWAPPED IN THE APPROXIMATE VALUES, WE CALL THE RESULT AN **ESTIMATED SAMPLING DISTRIBUTION:**

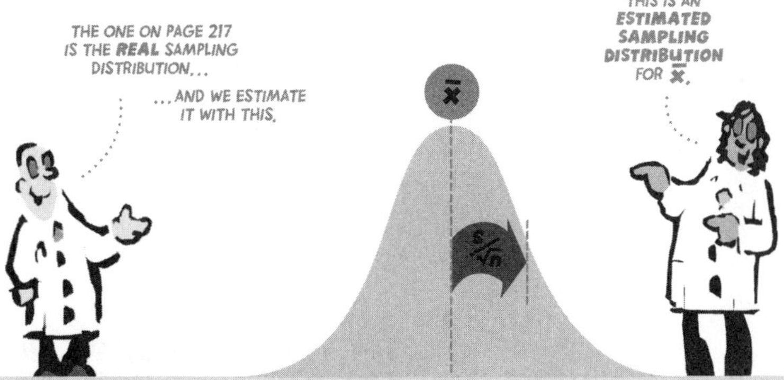
THE ONE ON PAGE 217 IS THE **REAL** SAMPLING DISTRIBUTION...
...AND WE ESTIMATE IT WITH THIS.
THIS IS AN **ESTIMATED SAMPLING DISTRIBUTION** FOR $\bar{x}$.
$\bar{x}$
$s/\sqrt{n}$
AN **ESTIMATED** DISTRIBUTION OF ALL POSSIBLE VALUES FOR $\bar{x}$

NOTE THAT WE CAN ALSO BUILD **ESTIMATED SAMPLING DISTRIBUTIONS** FOR OTHER STATISTICS, SUCH AS s (SEE PAGE **201**, FIRECRACKERS), BUT WE CAN ONLY EXPECT A SAMPLING DISTRIBUTION TO BE NORMAL-SHAPED WHEN THE CLT OR SIMILAR RESULTS APPLY.

FROM P. 135

CONFIDENCE INTERVALS

TECHNICALLY, A **CONFIDENCE INTERVAL** IS A TYPE OF INTERVAL ESTIMATE THAT RELATES TO A PARTICULAR **CONFIDENCE LEVEL**. CONFIDENCE INTERVALS CAN BE COMPUTED FOR ANY PARAMETER, ALTHOUGH THE SPECIFIC TECHNICAL DETAILS WILL CHANGE. HERE'S THE FORMULA FOR HOW TO COMPUTE A **95% CONFIDENCE INTERVAL** FOR A POPULATION AVERAGE μ :

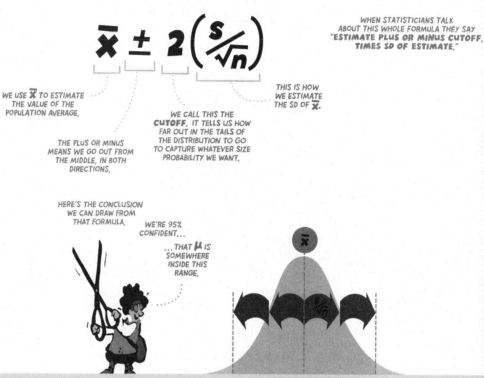

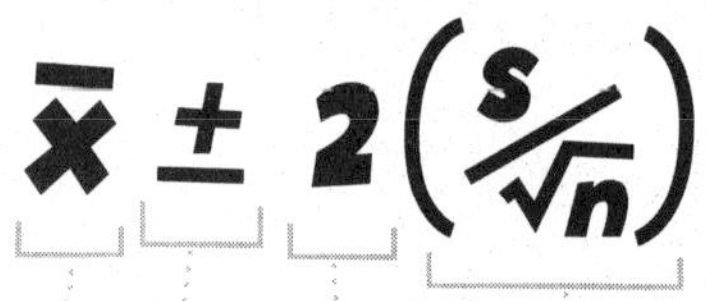

$$\bar{x} \pm 2\left(\frac{s}{\sqrt{n}}\right)$$

WHEN STATISTICIANS TALK ABOUT THIS WHOLE FORMULA THEY SAY "**ESTIMATE PLUS OR MINUS CUTOFF, TIMES SD OF ESTIMATE.**"
WE USE $\bar{x}$ TO ESTIMATE THE VALUE OF THE POPULATION AVERAGE.
THE PLUS OR MINUS MEANS WE GO OUT FROM THE MIDDLE, IN BOTH DIRECTIONS.
WE CALL THIS THE **CUTOFF**. IT TELLS US HOW FAR OUT IN THE TAILS OF THE DISTRIBUTION TO GO TO CAPTURE WHATEVER SIZE PROBABILITY WE WANT.
THIS IS HOW WE ESTIMATE THE SD OF $\bar{x}$.

HERE'S THE CONCLUSION WE CAN DRAW FROM THAT FORMULA.
WE'RE 95% CONFIDENT...
...THAT μ IS SOMEWHERE INSIDE THIS RANGE.
$\bar{x}$
$s/\sqrt{n}$

AN **ESTIMATED** SAMPLING DISTRIBUTION FOR $\bar{x}$

WE CAN **CHANGE OUR CONFIDENCE LEVEL** BY **CHANGING THE CUTOFF**. FOR EXAMPLE, IF WE WANTED AN **80%** CONFIDENCE INTERVAL FOR THE POPULATION AVERAGE, WE WOULD USE **1.3** AS OUR CUTOFF, SINCE APPROXIMATELY **80%** OF A NORMAL DISTRIBUTION IS CONTAINED WITHIN **1.3** STANDARD DEVIATIONS OF THE CENTER. (FOR AN EXAMPLE, SEE PAGE **157**.)

IDEALLY, WE WANT THE **NARROWEST POSSIBLE INTERVAL** FOR ANY LEVEL OF CONFIDENCE, SINCE A NARROWER INTERVAL IS MORE PRECISE. ONE SUREFIRE WAY TO GET A NARROWER INTERVAL IS TO INCREASE n (BY COLLECTING MORE OBSERVATIONS). THAT'S WHY A BIGGER SAMPLE SIZE IS BETTER! (FOR AN EXAMPLE, SEE PAGE **159**.)

REMEMBER THAT OUR **LEVEL OF CONFIDENCE** IS BASED ON A PROBABILITY VALUE, SO IT'S ONLY RELEVANT WHEN WE THINK ABOUT THE LONG RUN. AS A RESULT, WHEN WE COMPUTE AN INTERVAL USING THE FORMULA ABOVE, **WE DON'T KNOW WHETHER IT ACTUALLY CONTAINS μ OR NOT!** ALL WE CAN SAY IS THAT INTERVALS CONSTRUCTED IN THIS WAY WILL TEND TO BE ACCURATE IN THE LONG RUN. FOR A **95%** CONFIDENCE INTERVAL WE CAN EXPECT TO BE WRONG **5%** OF THE TIME... IN THE LONG RUN.

FROM P. 163

HYPOTHESIS TESTS

HYPOTHESIS TESTING USES THE **SAME UNDERLYING STATISTICAL MACHINERY** THAT WE USE WHEN WE COMPUTE A CONFIDENCE INTERVAL. WE STILL START BY BUILDING AN ESTIMATED SAMPLING DISTRIBUTION. THIS TIME, HOWEVER, WE USE IT TO QUESTION WHETHER WE THINK A **PARTICULAR VALUE** FOR THE POPULATION PARAMETER IS TRUE OR NOT. WE DO THIS BY ASKING **HOW CONSISTENT** OUR OBSERVED DATA ARE WITH THAT PARTICULAR VALUE.

FORMALLY, HYPOTHESIS TESTS START WITH TWO HYPOTHESES, ONE IS OUR **RESEARCH HYPOTHESIS** (SOMETIMES CALLED THE **ALTERNATE HYPOTHESIS)** AND THE OTHER IS THE **NULL HYPOTHESIS** (IN THE BOOK WE USE THE WORD "DULL").

HYPOTHESIS TESTS ALWAYS END WHEN WE CALCULATE A **P-VALUE** AND USE IT TO MAKE A **FORMAL DECISION** ABOUT WHETHER WE THINK OUR STATISTIC IS FAR ENOUGH AWAY FROM THE PARAMETER PREDICTED BY THE NULL HYPOTHESIS TO JUSTIFY REJECTING THE NULL HYPOTHESIS IN FAVOR OF ANOTHER EXPLANATION.

HERE IS A QUICK SUMMARY OF THE UNDERLYING LOGIC:

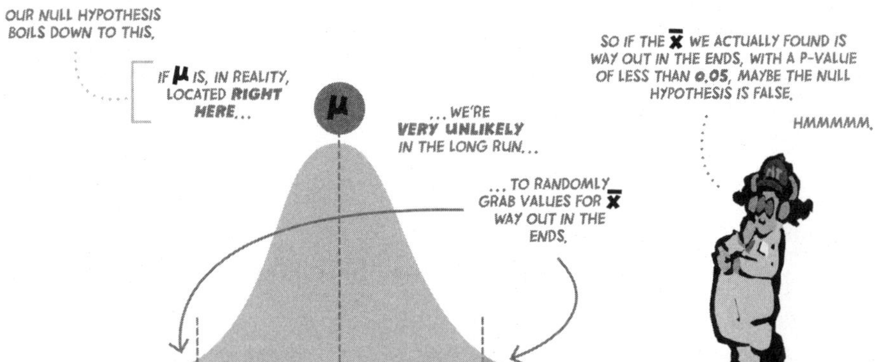
OUR NULL HYPOTHESIS BOILS DOWN TO THIS.
IF μ IS, IN REALITY, LOCATED **RIGHT HERE**...
μ
... WE'RE **VERY UNLIKELY** IN THE LONG RUN...
... TO RANDOMLY GRAB VALUES FOR $\bar{x}$ WAY OUT IN THE ENDS.
SO IF THE $\bar{x}$ WE ACTUALLY FOUND IS WAY OUT IN THE ENDS, WITH A P-VALUE OF LESS THAN **0.05**, MAYBE THE NULL HYPOTHESIS IS FALSE.
HMMMMM.

AN **ESTIMATED** SAMPLING DISTRIBUTION FOR $\bar{x}$, ASSUMING THE NULL HYPOTHESIS IS TRUE

*IN THIS BOOK WE'VE FOCUSED ON HYPOTHESIS TESTS ABOUT AVERAGES. IN PRACTICE, THESE SAME GENERAL STEPS CAN WORK FOR **ANY PARAMETER** AND ITS CORRESPONDING STATISTIC, BUT THE MATH DETAILS WILL VARY.*

FROM P. 169

P-VALUES

FORMALLY, A **P-VALUE** CAN BE DEFINED AS THE PROBABILITY THAT WE WOULD OBSERVE DATA AT LEAST AS EXTREME AS THE DATA WE ACTUALLY OBSERVED IF THE NULL HYPOTHESIS WERE TRUE. WHEW! IN THE BOOK WE INDICATE P-VALUES WITH PERCENTAGES, BUT AGAIN IT'S COMMON TO USE NUMBERS BETWEEN **0** AND **1**. A P-VALUE OF **5%** IS COMMONLY EXPRESSED BY THE NUMBER **0.05**.

SOMETIMES WE CALCULATE A P-VALUE FOR **BOTH ENDS** OF OUR ESTIMATED SAMPLING DISTRIBUTION (SEE PAGE **181**; THIS IS CALLED A **TWO-TAILED TEST**), AND SOMETIMES WE CALCULATE A P-VALUE FOR **ONE END** ONLY (SEE PAGE **187**; THIS IS CALLED A **ONE-TAILED TEST**). THE CHOICE OF WHICH TO USE DEPENDS ON WHAT SORT OF RESEARCH HYPOTHESIS WE'RE CURIOUS ABOUT.

IN PRACTICE, STATISTICIANS USE COMPUTERS TO **CALCULATE A P-VALUE** (CALCULUS AGAIN). HOWEVER, IT CAN BE HELPFUL TO NOTE THAT (WHEN WE'RE PERFORMING A TWO-TAILED TEST) A PROBABILITY OF **0.05** IS PRECISELY THAT AREA THAT **DOES NOT FIT** INSIDE A **95%** CONFIDENCE INTERVAL. AS A RESULT, ONE RELATIVELY SIMPLE WAY TO CARRY OUT A HYPOTHESIS TEST IS TO CONSTRUCT A 95% CONFIDENCE INTERVAL FOR μ AS DESCRIBED ABOVE. IF THE VALUE OF μ PREDICTED BY THE NULL HYPOTHESIS ISN'T INSIDE THAT INTERVAL, THE P-VALUE MUST BE LESS THAN **0.05**.

ON A RELATED NOTE: INCREASES IN n RESULT IN SMALLER P-VALUES. THAT'S WHY, IN STATS JARGON, COLLECTING MORE OBSERVATIONS IS A SUREFIRE WAY TO GET MORE **POWER** TO REJECT A NULL HYPOTHESIS. IT'S ANOTHER REASON WHY A BIGGER SAMPLE SIZE IS BETTER!

P-VALUES (CONT.)

REMEMBER THAT A P-VALUE IS A MEASURE OF PROBABILITY, SO IT'S ONLY RELEVANT WHEN WE THINK ABOUT THE LONG RUN.

IN PRACTICE, WE REJECT THE NULL HYPOTHESIS IF OUR P-VALUE IS "SUFFICIENTLY SMALL," WHICH (USING A COMMON RULE OF THUMB) MEANS **LESS THAN 0.05**, BUT THERE'S NOTHING MAGICAL ABOUT THAT NUMBER. A PROBABILITY OF "LESS THAN 0.05" MEANS THE SAME THING AS "FEWER THAN 1 OUT OF EVERY 20 TIMES IN THE LONG RUN."

SO FOR EXAMPLE, IF WE PERFORM A HYPOTHESIS TEST AND GET A P-VALUE OF **0.049**, IT MEANS THAT "IF THE NULL HYPOTHESIS WERE TRUE, WE'D EXPECT, JUST BY CHANCE, TO SEE DATA LIKE OURS ABOUT **49** OUT OF EVERY **1,000** TIMES IN THE LONG RUN." BECAUSE 49/1,000 IS LESS THAN 1/20, WE'D CONCLUDE THAT OUR DATA DON'T MATCH THAT NULL HYPOTHESIS WELL.

FROM P. 172

WE ALWAYS MIGHT BE WRONG

ALL STATISTICAL INQUIRY IS BASED ON RANDOM SAMPLING, AND **ALL** STATISTICAL INFERENCE IS BASED ON CALCULATING PROBABILITIES. AS A RESULT, ANY TIME WE USE A SAMPLE STATISTIC TO MAKE A GUESS ABOUT A POPULATION PARAMETER WE **MIGHT BE WRONG!**

BECAUSE OF THIS FACT, WE HAVE TO BE **VERY CAREFUL** ABOUT THE LANGUAGE WE USE WHEN WE'RE TEMPTED TO MAKE **TRUTH CLAIMS** BASED ON STATISTICS. WE HAVE TO BE ESPECIALLY CAREFUL WHEN WE'RE MAKING FORMAL CONCLUSIONS BASED ON P-VALUES, BECAUSE WE ONLY USE P-VALUES WHEN WE'RE INVESTIGATING THEORIES THAT WE'RE EXCITED ABOUT.

IF WE'RE INVESTIGATING A THEORY AND WE USE A SMALL P-VALUE TO ADD SUPPORT TO IT, WE **MIGHT BE WRONG**. OUR THEORY MIGHT BE WRONG AND CHANCE VARIATION MIGHT BE A BETTER EXPLANATION FOR OUR RESULTS. STATISTICIANS CALL THIS A **FALSE POSITIVE**, OR **TYPE 1** ERROR.

ALTERNATIVELY, IF WE'RE INVESTIGATING A THEORY AND WE USE A LARGER P-VALUE TO REJECT IT, WE **MIGHT BE WRONG**. OUR THEORY MIGHT ACTUALLY BE TRUE AND WE GOT RESULTS CLOSE TO THE VALUE PREDICTED BY THE NULL JUST BY CHANCE. STATISTICIANS CALL THIS A **FALSE NEGATIVE**, OR **TYPE 2** ERROR.

IN SUM, HYPOTHESIS TESTS ARE ONLY ABOUT ASKING THE QUESTION, "HOW LIKELY IS IT THAT WE JUST GOT OUR RESULTS BY CHANCE?" THEY CAN'T BE USED TO DISPROVE OR PROVE ANY THEORY CONCLUSIVELY; THEY CAN **ONLY** BE USED TO HELP US **CHALLENGE A NULL HYPOTHESIS**.

IN STATISTICS WE **ALWAYS MIGHT BE WRONG**. THIS IS ALWAYS THE CASE. IT'S A RESULT OF THE FACT THAT WE'RE USING A LONG-RUN PORTRAIT TO EVALUATE SHORT-TERM OBSERVATIONS.

FROM P. 197

INFERENCE ON A DIFFERENCE

TO CALCULATE A CONFIDENCE INTERVAL ABOUT THE **DIFFERENCE** BETWEEN TWO POPULATION AVERAGES, WE CAN USE A FORMULA THAT'S ONLY A BIT DIFFERENT FROM THE ONE WE LEARNED ABOVE.

IN THIS CASE, WE'RE CURIOUS ABOUT THE **DIFFERENCE** BETWEEN TWO POPULATION AVERAGES, AND WE ESTIMATE THAT DIFFERENCE WITH TWO SAMPLE AVERAGES.

THIS IS HOW WE **COMBINE THE VARIABILITY** OF THE TWO POPULATIONS.

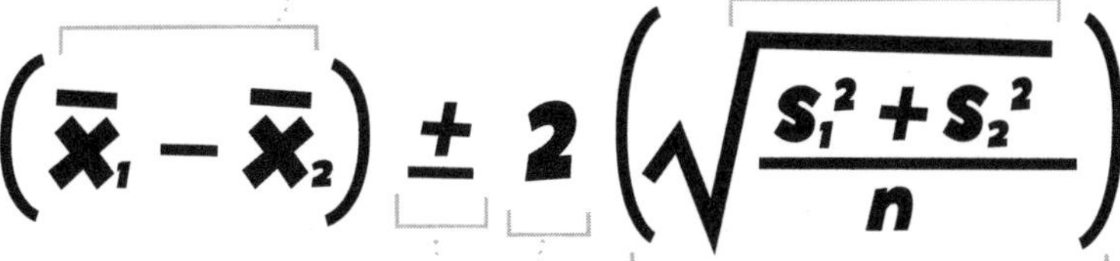

$$(\bar{x}_1 - \bar{x}_2) \pm 2\left(\sqrt{\frac{s_1^2 + s_2^2}{n}}\right)$$

THE PLUS OR MINUS MEANS WE GO OUT FROM THE MIDDLE, IN BOTH DIRECTIONS.

WE USE A **CUTOFF** OF 2 IF WE WANT A 95% CONFIDENCE INTERVAL, BUT OF COURSE THIS CAN BE CHANGED.

THIS WHOLE THING IS THE **SD OF OUR ESTIMATED SAMPLING DISTRIBUTION.** IN THIS STORY IT EQUALS APPROXIMATELY 1.

IN OUR STORY, $\bar{x}_1 = 59.7$, $s_1 = 4.6$, $\bar{x}_2 = 44.2$, AND $s_2 = 4.7$.

NOTE THAT THIS FORMULA CAN REQUIRE OTHER TWEAKS AS WELL. WE HAVE TO TWEAK IT, FOR EXAMPLE, IF WE HAVE **DIFFERENT SAMPLE SIZES FOR OUR TWO SAMPLES**, OR IF THEY'RE **TOO SMALL TO YIELD A NORMAL-SHAPED SAMPLING DISTRIBUTION.**

FROM P. 199

INFERENCE WITH A SMALL SAMPLE SIZE

WHENEVER WE'RE DOING INFERENCE ON A POPULATION AVERAGE AND WE HAVE A **SMALL SAMPLE SIZE** (E.G., WHEN $n<30$), WE CAN'T RELY ON THE **CLT**, SO WE USE WHAT'S CALLED THE **T-DISTRIBUTION**, WHICH WORKS **ONLY** IF THE POPULATION ITSELF IS NORMAL.

FOR MURKY HISTORICAL REASONS RELATED TO ITS CODISCOVERY BY A GUY AT THE GUINNESS BREWING COMPANY, THE **T-DISTRIBUTION** IS ALSO KNOWN AS THE "**STUDENTS DISTRIBUTION.**"

THE T-DISTRIBUTION IS FATTER THAN THE STANDARD NORMAL DISTRIBUTION, AND IN PRACTICE, WE MODIFY THE T-DISTRIBUTION DEPENDING ON THE SIZE OF n (A SMALLER n REQUIRES A FATTER T). AS A RESULT, WHEN WE USE THE T-DISTRIBUTION INSTEAD OF THE NORMAL DISTRIBUTION, OUR CONFIDENCE INTERVALS WILL BE WIDER, AND OUR P-VALUES WILL BE BIGGER. IN BOTH CASES, WE REQUIRE GREATER STATISTICAL EVIDENCE TO ACHIEVE THE SAME LEVEL OF CONFIDENCE. IT'S LIKE WE'RE BEING PENALIZED FOR THE SMALL SAMPLE SIZE.

IN OUR STORY, THE SPECIFIC T-DISTRIBUTION WE USED IS T_9, WHICH STATISTICIANS CALL THE "T WITH 9 [AS IN $n-1$] DEGREES OF FREEDOM."

HERE ARE THE PH MEASUREMENTS FROM OUR 10 RANDOM ALIENS: **2.09, 2.39, 1.32, 2.99, 2.62, 2.60, 2.45, 2.13, 2.27, 2.95.** FROM THESE NUMBERS WE CAN CALCULATE $\bar{x} = 2.38$, $s = 0.48$, AND $n = 10$, WHICH WE PLUG INTO THE FOLLOWING FORMULA TO GET A **95%** CONFIDENCE INTERVAL:

$$\bar{x} \pm 2.26\left(\frac{s}{\sqrt{n}}\right)$$

WE USE A DIFFERENT **CUTOFF** DEPENDING ON WHICH T-DISTRIBUTION WE'RE USING AND HOW MUCH CONFIDENCE WE WANT.

T-DISTRIBUTIONS LOOK ALMOST NORMAL, BUT THEY'RE FATTER OUT HERE.

FROM P. 201

INFERENCE ON A STANDARD DEVIATION

IN LITTLE SUZIE'S STORY, THE **CLT** CAN'T COME TO OUR RESCUE. THE SAMPLING DISTRIBUTION OF A STANDARD DEVIATION IS **NOT** GUARANTEED TO BE NORMAL, AND WE **CAN'T** COMPUTE A CONFIDENCE INTERVAL USING ANYTHING LIKE THE FORMULA ON PAGE **220** ABOVE. HOWEVER, **THE BASIC STEPS ARE THE SAME:** WE GENERATE A PARTICULAR KIND OF SAMPLING DISTRIBUTION (WITH SOME SNAZZY MATH) AND WE GENERATE A **95%** CONFIDENCE INTERVAL BY CALCULATING THE PROBABILITIES UNDERNEATH IT (WITH SOME OTHER SNAZZY MATH).

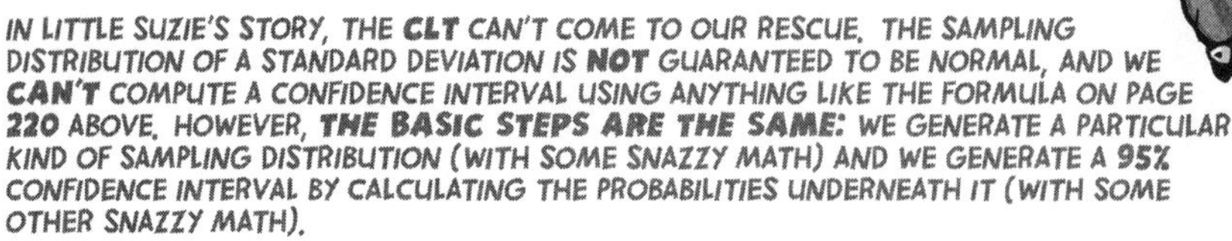

THE SNAZZY MATH IS TOO COMPLEX TO DESCRIBE HERE, BUT IT'S ALL GENERATED USING THESE FUSE TIMES FROM 15 RANDOM DINGALINGS: **2.05, 2.25, 2.33, 2.40, 1.66, 2.39, 1.89, 2.18, 2.18, 2.06, 2.18, 1.89, 2.14, 2.38, 2.07.** FROM THESE NUMBERS WE GET $\bar{x} = 2.14$, $s = 0.21$, AND $n = 15$.

NOTE THAT TO DO THIS KIND OF INFERENCE WE **HAVE TO ASSUME THE POPULATION IS NORMAL.** IN THIS CASE, IT MAY BE A FAIR ASSUMPTION, BUT IF THERE'S SOME KIND OF MANUFACTURING DEFECT THAT'S SKEWING THE OVERALL POPULATION (FOR EXAMPLE, AN UNEXPECTEDLY LARGE NUMBER OF FUSES FROM ONE FACTORY MIGHT BE DUDS, AND THIS ANALYSIS WOULDN'T ACCOUNT FOR THAT), OUR RESULTS **MIGHT BE MISLEADING.**

ALSO, LIKE THE T-DISTRIBUTION, THE SAMPLING DISTRIBUTION OF THE STANDARD DEVIATION CHANGES DEPENDING ON THE SAMPLE SIZE. AS IN ALL CASES, A LARGER n **WILL GIVE US MORE CONFIDENCE IN OUR CONCLUSIONS.**

FROM P. 202

CORRELATION

CORRELATION BEDEVILS LOTS OF STATISTICAL ANALYSIS. IF OUR SAMPLE MEASUREMENTS $x_1, x_2, x_3 \ldots x_n$ "CO-RELATE," THEY AREN'T INDEPENDENT, AND WE CAN'T USE ANY OF OUR STATS TOOLS ON THEM. SO WHENEVER DATA ARE CORRELATED, WE HAVE TO ACCOUNT FOR THE CORRELATION WHEN DOING INFERENCE.

FOR EXAMPLE, HEALTH STUDIES THAT INVOLVE **BIOLOGICAL TWINS** HAVE TO ACCOUNT FOR THE FACT THAT EACH TWIN'S DATA IS CORRELATED WITH THEIR SIBLING'S DATA, BUT WITH NO ONE ELSE'S. THIS CORRELATION CAN BE ELIMINATED USING WHAT'S CALLED A **PAIRED TEST.** OTHER CASES INVOLVE MEASUREMENTS THAT ARE CORRELATED **ACROSS GEOGRAPHICAL SPACE** (AS IN THE GECKO EXAMPLE ON PAGE **202**), OR **ACROSS TIME** (IMAGINE A CASE WHERE AN INDIVIDUAL ALIEN'S SALIVA pH VARIES DEPENDING ON THE TIME OF DAY AND WE OBTAIN MULTIPLE MEASUREMENTS OVER TIME FOR THE SAME ALIEN). WE CAN OFTEN INCORPORATE THIS TYPE OF CORRELATION BY ANTICIPATING IT IN LARGER MATHEMATICAL MODELS. IN GENERAL, EACH TYPE OF CORRELATION REQUIRES ITS OWN SEPARATE TRICK.

FROM P. 203

REGRESSION ANALYSIS

IN GENERAL, WE USE **REGRESSION ANALYSIS** WHEN WE WANT TO EXPLORE THE RELATIONSHIP BETWEEN A **RESPONSE** VARIABLE AND ONE OR MORE **PREDICTOR** VARIABLES. IN THE EXAMPLE ON PAGE **203**, SHRINKING MEDICINE DOSAGE IS THE PREDICTOR VARIABLE AND SHRINKAGE AMOUNT IS THE RESPONSE VARIABLE.

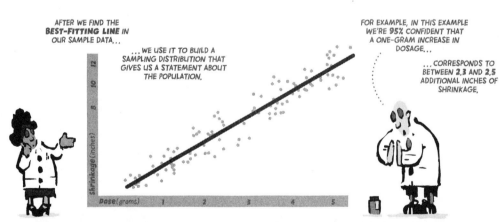

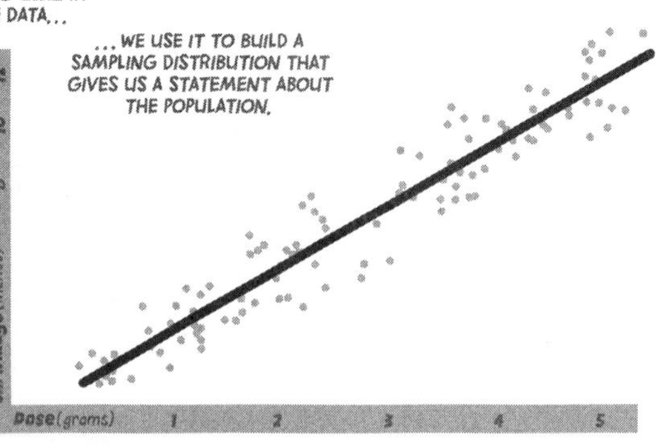
AFTER WE FIND THE **BEST-FITTING LINE** IN OUR SAMPLE DATA...
...WE USE IT TO BUILD A SAMPLING DISTRIBUTION THAT GIVES US A STATEMENT ABOUT THE POPULATION.
Shrinkage (inches)
8
10
12
Dose (grams)
1
2
3
4
5

FOR EXAMPLE, IN THIS EXAMPLE WE'RE **95%** CONFIDENT THAT A ONE-GRAM INCREASE IN DOSAGE...
...CORRESPONDS TO BETWEEN **2.3** AND **2.5** ADDITIONAL INCHES OF SHRINKAGE.

IN PRACTICE, WE HAVE TO BE CAREFUL WHEN WE'RE CURIOUS AS TO WHETHER PREDICTOR VARIABLES ACTUALLY **CAUSE** CHANGES IN RESPONSE VARIABLES. CAUSATION REQUIRES VERY CAREFUL EXPERIMENTAL DESIGN.

WE CAN ALSO USE THIS TYPE OF ANALYSIS TO COMPARE QUALITIES LIKE THE **HEIGHT** AND **WEIGHT** IN A POPULATION. IN THAT CASE, THE ESTIMATED SLOPE OF THE LINE WOULD TELL US ABOUT THE AVERAGE CHANGE IN WEIGHT FOR EVERY ONE-INCH INCREASE IN HEIGHT.

FROM P. 203

ANOVA

ANALYSIS OF VARIANCE (ANOVA) IS A HYPOTHESIS TESTING TECHNIQUE, BUT IT'S VERY DIFFERENT FROM THE HYPOTHESIS TESTING WE'VE LEARNED IN THIS BOOK. ANOVA WORKS BY COMPARING THE **VARIABILITY BETWEEN GROUPS** TO THAT **WITHIN GROUPS**. WE CAN USE IT WHEN WE WANT TO COMPARE **MORE THAN TWO** SAMPLE AVERAGES. THERE ARE **LOTS** OF WAYS TO USE ANOVA, AND HERE'S A QUICK RUNDOWN OF ONE:

I WANNA KNOW, DO THESE 5 SEPARATE FIRECRACKER SAMPLES ALL COME FROM A POPULATION WITH THE SAME AVERAGE...
...MAYBE FROM THE SAME FACTORY?

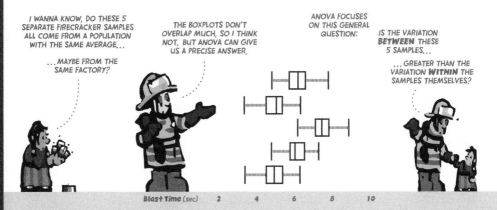

THE BOXPLOTS DON'T OVERLAP MUCH, SO I THINK NOT. BUT ANOVA CAN GIVE US A PRECISE ANSWER.

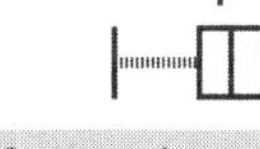
Blast Time (sec)
2
4
6
8
10

ANOVA FOCUSES ON THIS GENERAL QUESTION:
IS THE VARIATION **BETWEEN** THESE 5 SAMPLES...
...GREATER THAN THE VARIATION **WITHIN** THE SAMPLES THEMSELVES?

FROM P. 203

INFERENCE ON A PROPORTION

IF WE'RE CURIOUS ABOUT **WHAT PROPORTION** OF FOOTBALL FANS PREFER CHEESE DOODLES OVER PORK RINDS, OR **WHAT PERCENTAGE** OF LIKELY VOTERS PLAN TO CHOOSE TO RE-ELECT SENATOR SAM WARM IN THE UPCOMING ELECTION, WE CAN OFTEN USE THE BASIC INFERENCE STEPS WE'VE LEARNED IN THIS BOOK, BUT WE HAVE TO TWEAK THE DETAILS.

FOR EXAMPLE, HERE'S HOW WE COMPUTE A 95% CONFIDENCE INTERVAL FOR A POPULATION PROPORTION p.

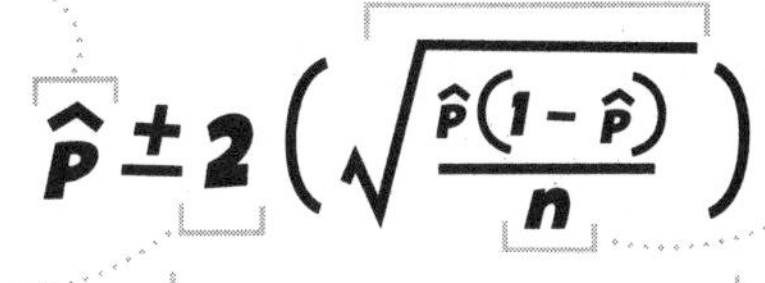

$$\hat{p} \pm 2\left(\sqrt{\frac{\hat{p}(1-\hat{p})}{n}}\right)$$

WE USE OUR SAMPLE PROPORTION $\hat{p}$ (P-HAT) TO ESTIMATE THE POPULATION PROPORTION.
THIS IS THE STANDARD DEVIATION OF A SAMPLE PROPORTION.
WITH A LARGE SAMPLE SIZE, WE CAN USE A STANDARD NORMAL SAMPLING DISTRIBUTION, SO TO GET A 95% PROBABILITY, WE USE A CUTOFF OF 2 HERE.
ONCE AGAIN, WE NEED A LARGE SAMPLE SIZE TO MAKE IT WORK, AND THE LARGER THE SAMPLE SIZE, THE NARROWER OUR INTERVAL!
IN OPINION POLLING WE REFER TO THIS AS THE **MARGIN OF ERROR**.

FROM P. 203

PREDICTING THE FUTURE

IN THIS BOOK WE'VE FOCUSED ON USING OUR SAMPLE DATA TO CHARACTERIZE ASPECTS OF AN OVERALL POPULATION, ITS AVERAGE, SAY, OR ITS STANDARD DEVIATION. HOWEVER, WE CAN ALSO USE STATISTICAL INFERENCE TO MAKE PREDICTIONS ABOUT **SINGLE OBSERVATIONS**. FOR EXAMPLE, WE CAN ASK QUESTIONS LIKE "BASED ON THE MEASUREMENTS I'VE GOT, WHAT'S THE **NEXT MEASUREMENT** (X_{n+1}) PROBABLY GOING TO BE?"

ONCE AGAIN WE CAN MAKE SOME HEADWAY USING OUR BASIC INFERENCE STEPS. FOR EXAMPLE, IF WE ASSUME THE OVERALL POPULATION IS NORMAL (ALWAYS A DICEY ASSUMPTION), WE CAN CREATE A "**PREDICTION INTERVAL**" THAT'S SIMILAR TO A STANDARD CONFIDENCE INTERVAL BUT A LITTLE BIT WIDER.

IN PRACTICE, STATISTICIANS DO THIS SORT OF THING TO PREDICT WEATHER PATTERNS OR FUTURE PRICES IN FINANCIAL MARKETS, THOUGH THEY USE MORE COMPLEX TOOLS.

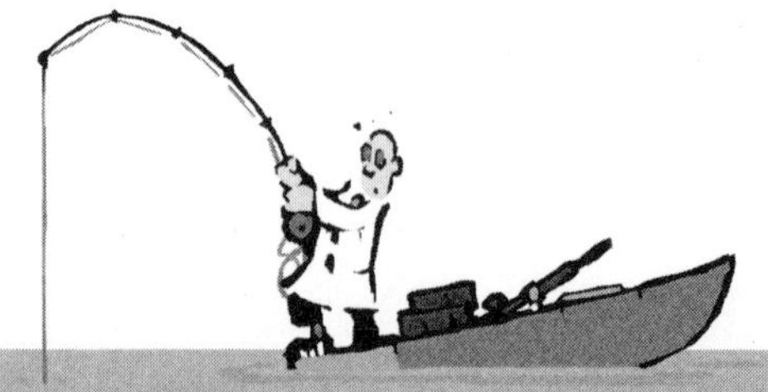

A NOTE ABOUT THE AUTHORS

GRADY KLEIN IS A CARTOONIST, ILLUSTRATOR, AND ANIMATOR. HE IS THE COAUTHOR OF *THE CARTOON INTRODUCTION TO ECONOMICS*, VOLUMES ONE AND TWO, AND THE CREATOR OF THE *LOST COLONY* SERIES OF GRAPHIC NOVELS. HE LIVES IN PRINCETON, NEW JERSEY, WITH HIS WIFE AND TWO CHILDREN.

ALAN DABNEY, PH. D., IS AN AWARD-WINNING ASSOCIATE PROFESSOR OF STATISTICS AT TEXAS A&M UNIVERSITY. HE LIVES IN COLLEGE STATION, TEXAS, WITH HIS WIFE AND THREE CHILDREN.